The Miegunyah Press

This is number one hundred and sixty in the
second numbered series of the
Miegunyah Volumes
made possible by the
Miegunyah Fund
established by bequests
under the wills of
Sir Russell and Lady Grimwade.

'Miegunyah' was the home of
Mab and Russell Grimwade
from 1911 to 1955.

Transforming Biology

A History of the
Department of Biochemistry
and Molecular Biology
at the University of Melbourne

JULIET FLESCH

THE
MIEGUNYAH
PRESS

THE MIEGUNYAH PRESS
An imprint of Melbourne University Publishing Limited
11–15 Argyle Place South, Carlton, Victoria 3053, Australia
mup-info@unimelb.edu.au
www.mup.com.au

First published 2015
Text © Juliet Flesch, 2015
Design and typography © Melbourne University Publishing Limited, 2015

National Library of Australia Cataloguing-in-Publication entry

Flesch, Juliet, author.

Transforming biology: a history of the Department of
Biochemistry and Molecular Biology at the
University of Melbourne/Juliet Flesch.

9780522867701 (hardback)

Includes bibliographical references and index.

University of Melbourne. Department of Biochemistry and
Molecular Biology—History.
Universities and colleges—Victoria—Melbourne—History.

378.9451

Cover design by Nada Backovic
Typeset in Minion 11/15pt by Cannon Typesetting
Printed in China by 1010 Printing International

This book is dedicated to the memory of
Dr Valda McRae (1935–2014),
an exceptional university administrator
and even more exceptional friend

and to

Emeritus Professor Francis Hird (1920–2014),
a major biochemist and inspirational guide to many.

Foreword

IT WAS WITH great pleasure that I accepted Professor Paul Gleeson's invitation to write a short foreword to *Transforming Biology*, the first history of the Department of Biochemistry and Molecular Biology at the University of Melbourne, written on the occasion of its 75th anniversary by Juliet Flesch.

My first encounter with the department was during festivities in Orientation Week in 1962. During a barbeque, I was approached by two women who introduced themselves as lecturers from the Biochemistry Department – Dr Mary McQuillan and Dr Pam Todd. When she heard my name, Dr McQuillan eagerly asked if I was related to the Nobel Prize– winning biochemists Gerty and Carl Cori. I was not, of course, and at that stage I had no idea what biochemistry was! But the unusual conversation stuck in my mind and probably subconsciously influenced my decision to take biochemistry the next year.

Like so many other undergraduates who passed through its doors, I received a marvellous education in biochemistry at the University of Melbourne. I have vivid memories of the lectures by Michael Birt, Lloyd Finch, Cai Mauritzen, Frank Hird, Bruce Stone and Jack Legge – such dif- ferent personalities but each so passionate about their fields. Biochemistry was undergoing a major revolution at the time. While Hans Krebs and the TCA cycle still reigned supreme – especially in the prac lab, where we struggled with the intricacies of the Warburg apparatus – molecular biology was starting to transform understanding of the life process. Enthralled, I heard the story of Watson and Crick's discovery of DNA and learned that the genetic code had just been cracked. After I finished my BSc, Michael Birt asked me if I was interested in doing a research degree. My future plans were very vague at that stage, so I agreed to meet him the next day and eventually enrolled for an MSc. It transpired that Michael was headhunted to Sheffield University and so Lloyd Finch, who shared a lab with him, inherited me as a student. Although I began with trepidation, I found that I enjoyed lab life and relished being able to work on nucleic acids. I became an eager member of the molecular genetics literature club Lloyd ran with Jim Pittard and Bruce Holloway from the adjacent Microbiology Department. These formative years eventually led me to PhD studies in Cambridge, in the famed MRC Laboratory of Molecular Biology, which set the course for my subsequent career.

I was not alone in being advantaged by my training in the Biochemistry Department at the University of Melbourne. As Dr Flesch amply documents, its graduates were eagerly accepted as postgraduate trainees in major laboratories around the world. Many subsequently returned to make major contributions to Australian science. Others became expatriate stars, but their links to the department remained strong and opened doors for subsequent generations of students and postdoctoral fellows.

Dr Flesch surveys the early history of biochemistry within the Physiology Department; its birth as an independent department under Professor William John Young; its rise to prominence under the amazing 25-year 'reign' of Professor Victor Trikojus; its growing embrace of molecular biology from the 70s; and its productive links with other departments, CSIRO and the Walter and Eliza Hall Institute. She brings to life the many scientists and teachers who made the department so successful, noting the high proportion of women at a time when this was still rare.

She also pays tribute to the Grimwade family, whose remarkable philanthropy enabled the department to build modern, purpose-built facilities on Royal Parade in three stages that were opened in 1958, 1961 and 1966. But times move on. In 2005, to catalyse greater convergence between disciplines, the Department of Biochemistry and Molecular Biology relocated its research laboratories to the new Bio21 Institute, where they have continued to flourish. And in 2008, nostalgically watched by many of us who had trained or taught there, the Russell Grimwade School was razed to make way for the new neurosciences institute. Juliet Flesch has ensured, however, that the 50 years of teaching and research nurtured there will not be forgotten.

I commend *Transforming Biology* to you. Its title reminds us how the molecular biosciences are transforming understanding of evolution and the life process and how this knowledge can be harnessed for improving health, agriculture and industry. The first 75 years of the Department of Biochemistry and Molecular Biology at the University of Melbourne was a period of impressive growth and achievement and I have every confidence that trajectory will continue.

Suzanne Cory
30 September 2014

Contents

Introduction

THE HISTORY OF any department of the University of Melbourne often reads, after more than a century and a half, as a more or less constant litany of complaints about inadequate accommodation, briefly and occasionally interrupted by celebrations of new buildings being opened or old ones refitted and refurbished. It is rare, however, for a transformation in a particular department to be almost entirely due to the generosity and longstanding support of a single family. In the history of Biochemistry at the University of Melbourne, the support of Russell Grimwade and his family is paramount.

Although the study of biochemistry at the university predated the establishment of the department by many years, and the department itself was established six years before Russell Grimwade announced his transformative support, it is appropriate to begin this history of the first seventy-five years of Biochemistry at Melbourne with some account of the relationship between Grimwade and the university. A fuller account of the life of Russell Grimwade can be found in the biography by J.R. Poynter and in Chapter 5.[1]

Support for the University of Melbourne by the Grimwade family was neither initiated by Russell Grimwade nor confined to biochemistry. It came from many family members and took the form of gifts in kind as well as gifts of money. Although it falls outside the scope of our history, it is also worth noting that Russell Grimwade was a very significant benefactor of the School of Forestry in Canberra, and in 1929 he endowed the associated Grimwade Prize, which is awarded every three years. It is to him that Melbourne owes the presence of Captain Cook's Cottage in the Fitzroy Gardens. Melbourne Girls' Grammar benefitted greatly from the support of Norton Grimwade (Russell's eldest brother) and his wife Phelia. Three Grimwade brothers were longstanding and important benefactors of the National Gallery of Victoria: Harold, Norton and Russell all served on the board of trustees of the Felton Bequest.[2]

The first recorded Grimwade benefaction to the University of Melbourne came in 1905. The *Calendar* notes the gift from Russell Grimwade's father, Frederick Sheppard Grimwade, of £1000 for a Prize of Technical Chemical Research. Renamed the Grimwade Prize for Industrial Chemistry, it has been awarded (to biochemists and others) most years since 1907. A list of recipients can be found at the end of this book. In 1920 several benefactions from the family were made, notably £250 from Russell and another £250 from his brother Harold William Grimwade to the University Appeal.

That year also saw the first of many gifts from the family to Trinity College, with the Risdon Grimwade Lectureship in Chemistry established through a donation by Norton and Phelia Maud Grimwade in memory of their son George, a Trinity man who fell at Gallipoli aged twenty while serving in the Field Ambulance, Australian Army Medical Corps. In 1923 the Harold Grimwade Lectureship was established thanks to Harold William's payment 'of a stipend of a Lectureship of the College in Natural Philosophy (or in some other branch of Natural Science)'.[3] In 1955 Harold Thornton Grimwade, Russell's nephew, gave £100 to the University Centenary Appeal.[4]

The years 1933 to 1939 saw several donations from Russell Grimwade to the university, including £100 to the Orchestra Fund, £1000 for university grounds improvement, £1000 to the Engineering School Appeal and £1000 for meteorology research as well as the commissioning of Percival St John's survey of the flora of the Mount Buffalo region.[5] (The removal of the material St John collected to the Botany School herbarium in 2006 was funded by the Miegunyah Foundation.[6]) The war years saw Russell Grimwade give £1000 to the Chemistry School, £2000 to the Wilson Hall Fund, £2000 towards the furnishing of University House, £5000 to the Centenary Appeal and £100 towards a 'rural survey' in 1941. The 1950s saw donations of £100 to the International House Appeal and £400 to the Chair of Pharmacology.

It was, however, in 1944 that Russell Grimwade made his most spectacular offer of assistance to the University of Melbourne. It was to transform the study of biochemistry at the university and in the world at large. The story of the gift of £50 000 for the foundation of the Russell Grimwade School of Biochemistry forms the basis of Chapter 5 of this book.

Support from the Russell and Mab Grimwade Bequests and the Russell and Mab Grimwade Miegunyah Fund continue to benefit the university at large to this day. The Miegunyah Press, an imprint of what is now Melbourne University Publishing, began publishing in 1967. After Mab Grimwade's death, the university recorded in 1973 the receipt of $75 000 from the estate of Russell Grimwade for the use of Melbourne University Press and towards the Department of Fine Arts. It also acknowledged gifts valued in excess of $1.2 million. These included extremely valuable land, the house the couple had lived in for all their married life, and its contents, as well as Russell Grimwade's collection of Australiana, consisting of books, pictures and prints. In 1975 a further $24 983 was received from the estate of Russell Grimwade for various departments within the university and 1990 saw the establishment of the Sir Russell and Lady Mab Grimwade Miegunyah Fund.[7] Following the sale in 2004 of 'Miegunyah', the house Mab Grimwade had lived in since 1911, the university received $8 810 710.52 from the Russell Grimwade Trust and $1 738 675.88 from the Mab Grimwade Trust. Nine per cent of the income is distributed annually. The Miegunyah Distinguished Visiting Fellowship Program has brought scholars to Melbourne since 2005 – it has provided funding of $20 000 each for five fellows in 2014.[8]

In 1989, support from the Ian Potter Foundation and the Sir Russell and Lady Mab Grimwade Miegunyah Fund enabled the establishment of a Centre for Cultural Materials Conservation. In 2014 a gift of almost $7 million from the Cripps Foundation enabled the university to establish a Chair of Cultural Materials Conservation as well as new laboratories in new premises, to be known as the Grimwade Centre for Art Conservation.[9]

A bare recital of financial support risks overshadowing the immeasurable cultural impact of these gifts on the university and the community at large. The various collections of books, pictures and so on are extraordinarily rich. They inspired several exhibitions between 1987 and 2000: the catalogues may be seen in the University Library.[10] Russell Grimwade was an accomplished photographer and in 2003, 600 photographs from his astonishing collection were digitised for viewing online. The Miegunyah Trust funded the first University of Melbourne Cultural Treasures Day in

2008. This event is now held in conjunction with the Australian & New Zealand Association of Antiquarian Booksellers Book Fair and provides an opportunity for the general public to view parts of the university's special collections. The same trust has enabled the cataloguing of books and prints in the Baillieu Library Special Collections, rare medical books and journals, rare earth sciences and East Asian books, historic maps, and eucalyptus and early specimens in the herbarium. It has funded the upgrading of the collection database for the Henry Forman Atkinson Dental Museum and the conservation of historic dental drawings; it has also enabled the condition surveying and conservation of scientific instruments in the Physics Museum, and the production of digital preservation copies of cassette tapes in the University of Melbourne Archives between 2007 and 2009. Finally, the very publication of this history of the department has been most generously supported by the Miegunyah Fund.

As we have noted, however, this book is concerned with the history of teaching and research in biochemistry and molecular biology at the University of Melbourne both before and after the Grimwades began supporting them, and it is to this that we will now turn.

Despite the problems occasioned by having to share facilities with other space-hungry departments, teach across multiple venues, and cope with architects and builders not always fully attuned to the department's needs, as well as the necessity of raising funds from private, corporate and government sources, staff in the Department of Biochemistry and Molecular Biology at Melbourne have nonetheless taught men and women who went on to achievements in Australia (including in their alma mater) and overseas, as well as conducting their own groundbreaking research.

The department has existed in its own right only since 1938. This account will concentrate on the first seventy-five years of its history, but since biochemists at the University of Melbourne were making their mark nationwide and overseas well before that time, our first chapter tells that story.

The department was born into troubled times and the death of its foundation professor in the middle of World War II left it leaderless for twelve months until his successor arrived from Sydney. The second chapter gives an account of the career of Victor Martin Trikojus, who was to lead the department from 1943 to 1968, and covers investigations undertaken within the department during the war as well as some of the difficulties with which staff had to contend.

The war years and those immediately following were remarkable for a strong female cohort in the department, many of whom were to have a lasting effect on the teaching of and research in biochemistry. Several were quite remarkably long-lived. All had interesting careers outside the department they did so much to shape. The third chapter tells some of their stories.

The fourth chapter is one of development and expansion leading to 1958 when the university finally saw the opening of the first stage of the Russell Grimwade School of Biochemistry in the new building that had been mooted almost a decade and a half earlier. We will read of Grimwade's first approaches to the university administration. 'The Great Shift' begins with a brief account of the life and work of Sir Russell himself. We will meet some of the men and women who experienced not only the long-awaited move into a dedicated building for biochemistry at the University of Melbourne, but also the seismic shift in their discipline that followed the discovery of the structure of DNA. Chapter 6 is the story of the Russell Grimwade Building itself, a building which was the result of Victor Trikojus's determination to bring his empire under one roof – and which was demolished just half a century later. It also records the naming of laboratories on the second floor of the South Tower of the Bio21 Institute as the Grimwade Research Laboratories.

Chapter 7 covers the careers of some of the many men and women who began their biochemistry investigations in the department but completed them elsewhere; it also looks at a handful who travelled in the other direction, crowning stellar careers with work at the University of Melbourne. Chapter 8 is devoted to the large number of people, often unacknowledged and, as far as the outside world is concerned, unknown, whose work makes that of the scientists possible.

Chapter 9 describes the work of the teaching and technical staff who remained on the Parkville campus when the research activity of the department shifted to the Bio21 Institute, as well as the work of the Grimwade research fellows, funded by bequests from Russell and Mab Grimwade. Chapter 10 covers some of the work of those who made the move in 2005 or joined the staff before 2013. The final chapter gives some indication of the direction the department expects to take as it continues a proud tradition of discovery and of collaboration with colleagues in other institutions and industry, as well as teaching.

Chemical Physiology Classroom, 1896

CHAPTER 1

In the Beginning

WILLIAM JOHN YOUNG, the foundation Professor of the Department of Biochemistry at the University of Melbourne, had been working in the Department of Physiology for almost two decades before his own department was established in 1938. Biochemistry, which is of central importance to the study of health and disease, and so essential to medicine, agriculture, botany, zoology and many other disciplines, had already been the subject of formal teaching and investigation at the university for over three-quarters of a century. A great deal of work was performed by biochemists at the university before William Alexander Osborne, Professor of Physiology, won his long battle to have a separate Department of Biochemistry set up.

The School of Medicine was established in 1862, with George Britton Halford (1824–1910) taking up his appointment as the foundation Professor of Anatomy, Physiology and Pathology. John Macadam (1827–1865) had begun giving lectures in chemistry to medical students the year before.[1] Because the university did not yet have any laboratory facilities, lectures, practical classes and examinations were held in Macadam's city laboratory in Russell Street: the first examination in practical chemistry took place in October 1862.[2] At the time, and until about 1900, biochemistry was commonly referred to as chemical physiology, as, for instance,

George Britton Halford,
first Professor of Anatomy,
Physiology and Pathology

John Macadam, parliamentarian,
public servant and lecturer in
theoretical and practical chemistry

in the well-known photograph of the chemical physiology class of 1896, which was taken from the back of the room so that the half-dozen female students, actually seated at the front of the class, appear at the back of the image. Despite the best efforts of Professor Halford, women had not been admitted to the medical course until 1887.

Macadam, after whom the macadamia nut is named, completed medical studies in Glasgow and was awarded an MD ad eundem statum from Melbourne in 1857. During his ten years in Australia he managed to occupy, with distinction, an extraordinary number of positions.[3] He was appointed from Glasgow in 1855 to teach chemistry and natural science at Scotch College, and while continuing to do so until his death, he was simultaneously the Victorian Government analyst from 1858, health officer to the City of Melbourne from 1860 and a member of the Legislative Assembly from 1859 to 1864. He was honorary secretary of the Philosophical Institute of Victoria from 1857 to 1859, editing its *Transactions* from 1855 to 1860. He was honorary secretary of the Royal Society when it succeeded the Philosophical Institute and its vice-president from 1863. As secretary of the Exploration Committee of the expedition of Burke and Wills, he ensured that its provisions were adequate, managing moreover to convince a public meeting convened to censure the committee that it had in no way contributed to the expedition's tragic outcome. He was appointed lecturer in theoretical and practical chemistry at the University of Melbourne in 1862.

Perhaps unsurprisingly, this immense workload took its toll on his health and Macadam was already unwell when he sailed to New Zealand to give evidence at the murder trial of Captain W.A. Jarvey, who had been charged with poisoning his wife. The jury failed to reach a verdict, and Macadam suffered pleurisy after a fall during the return voyage. He was still unwell when he set sail again for the resumption of the trial, accompanied by the man who would succeed him at the university, and died at sea on 2 September 1865. The *Australian Medical Journal*, while noting that he had never actually practised medicine, paid tribute to his other accomplishments, noting that:

> In general scientific attainments, Dr Macadam had few equals: in the department of chemistry, he had no equals in this colony. As a lecturer he possessed a peculiar facility in communicating knowledge; he was fluent in language, neat in manipulation, and skilful, and expert in conducting experiments.[4]

His funeral cortege of several dozen carriages included the Chancellor, Vice-Chancellor and Professor Halford together with other academics and medical students in academic dress as well as members of parliament, the judiciary and the medical profession. Macadam Street, in the Canberra suburb of Page, was gazetted in 1969 in his honour.

The young medical student who accompanied John Macadam on his ill-fated voyage to New Zealand, and gave evidence at the trial, returned to Melbourne and suspended his own studies to fill in for his former lecturer at the request of his fellow students. Russell notes that this was the first and possibly the only occasion on which students have requested the selection of a lecturer.[5] John Drummond Kirkland (c. 1836–1885) took his MB in 1873 and his BSc in 1880. He performed so well that he was offered a permanent position the following year, and to support his teaching the university bought all of Macadam's privately owned apparatus and materials from his widow.[6] However, already burdened with the salary of its first medical professor, the university struck a hard bargain. Having paid £262 for Macadam's material, it proved resistant to many later requests for laboratory equipment. Kirkland remained a lecturer until 1882, when he was appointed to the foundation Chair of Chemistry at a somewhat farcical meeting of the Melbourne University Council, in the course of which no fewer than five new professors were appointed, without overseas advertisement, from among the university's existing staff.[7]

Kirkland's career in the university has been traced by Radford.[8] His lasting achievement may have been finally convincing the council of the utter inadequacy of the laboratory accommodation: part of the £10 000 grant for a new Medical School in 1885 was to be used for the purpose, as can be seen in the 1896 photograph. He was succeeded in 1886 by David Orme Masson, at which time chemistry moved from the Medical School to become part of the science degree.

Until the arrival of a professor to replace Halford's successor Charles Martin in physiology, the teaching of biochemistry appears to have been shared between Masson, John Booth Kirkland (the son of John Drummond Kirkland) and Professor Halford, who is listed in the 1887 *Calendar* as lecturing in physiological chemistry and histology. Some biochemistry would also have been taught as part of the Materia Medica course taken by medical students in their second year.

From the time of his appointment as Professor of Physiology succeeding Charles Martin in 1903, William Alexander Osborne, himself the author of a book on the subject, had campaigned for a greater emphasis on biochemistry.[9] In 1905 he was successful and the first lectureship dedicated to the subject at an Australian university was approved. This had been greatly aided by the establishment the year before of the Faculty of Agriculture, and the council made clear from the outset that the lecturer would be expected to develop strong links with the agricultural industry. The conditions of appointment specified that:

1 The Lecturer will be required to devote the whole of his time to the work of the University.
2 The Lecturer will have independent control of the teaching of Students in Bio-Chemistry in the course of Agriculture. He will be required to deliver such Lectures and conduct such Laboratory

work and Examinations in this subject as the Council may from time to time appoint on the recommendation of the Faculty of Agriculture.

3   He will also act as Demonstrator of Physiological Chemistry for the Students in Medicine and Science, and in other courses in which the subject may be taught. With regard to this part of his work he will be subject to the general control of the Professor of Physiology.

4   The appointment will be for a period of five years subject to good behaviour and to the work being performed to the satisfaction of the Council to the end of that time the Lecturer will be eligible for reappointment.

5   The Lecturer will be a member of the Faculty of Agriculture.

6   The Salary will be £600 per annum.

As at present arranged, the Lectures will occupy about two hours a week during 27 weeks; the Laboratory work about 10 hours a week for the same period. The Laboratory (which is in the Department of Physiology and is used by arrangement with the Professor) has accommodation for 60 students. In addition, private rooms and a research laboratory will be provided for the Lecturer.[10]

John Drummond Kirkland, first
Chair of Chemistry

William Alexander Osborne,
Professor of Physiology, 1903–38

The academic requirements were also clarified:

It is expected that the Candidate chosen should have a sound knowledge of Organic Chemistry, and should have had some experience of work in a Physiological or Bio-Chemical Laboratory. While conversant with ordinary chemical technique and especially with the detection, preparation synthesis and estimation of the more common carbon compounds, it is also expected that he should have some knowledge of chemical method as applied to physiological work, such as the separation of organic compounds from tissues, the changes which these bodies undergo, the method of their elaboration and the functions which they perform in the living animal. As far as the teaching of Agricultural Science is concerned, it is desired that the general treatment of Bio-Chemistry shall be such as to lay a scientific basis for further work in Agricultural Botany and the Physiology and Bacteriology of domesticated animals.[11]

The eventual appointee, who would, had he survived, in all probability have headed the Department of Biochemistry when it was established in 1938, held his position for less than a decade, but the impact of his work was felt Australia-wide.[12] The name of Arthur Cecil Hamel Rothera (1880–1915) was not the first put forward for consideration for the post, with the council debating appointing either James Matthew Petrie (1872–1927) or Sydney Francis Ashby (1874–1954) as late as June 1906. This meant that once the formal offer was made, the young man had to move very fast indeed to get from England to Melbourne in time to begin lectures in August. Gowland Hopkins, who had supported Rothera's application for the position, described his situation:

The appointment came somewhat unexpectedly, and he was left with a very brief period – I believe it was a fortnight only – to wind up his affairs in this country, to get his equipment together, to sit for the final examination of the Conjoint Board, and, above all, to cross the Channel and persuade the family of the charming French lady who was to be his wife to consent to an arrangement which would later on involve her joining her husband in a country most remote. All was successfully accomplished in the brief period mentioned and Rothera sailed to begin a career which though it was to prove lamentably short was to bring him satisfaction and happiness.[13]

Hopkins describes Rothera's decision to take a position in Australia as a sign of 'true grit', and certainly both he and his young wife showed a good deal of courage and enterprise in venturing so far from home at a very young age: Rothera was only twenty-five when he took up the post; his wife, Rosalie Désirée Held (1882–1971), was two years younger. He made his mark early. A new building had already been erected in 1905–06 connected to the Old Medical Building in Madeline Street, with biochemical teaching and experimental physiology accommodated on the second floor. Rothera succeeded in getting additional accommodation in 1912, although, as Selleck tells us, the end result was not universally approved of because the additions involved the destruction of the portico of the building.[14] He joined the Melbourne University Chemical Society as soon as he arrived, participated in two of its debates in 1907 and delivered papers in 1909 and 1910.[15]

Arthur Rothera, the first lecturer in Biochemistry, with his son Jack in 1912

Arthur Rothera had already published his first paper in 1904, but the one that appeared two years after his arrival in Australia made his name.[16] The Rothera Test for diabetes is still used and the paper was still being cited a century after publication.[17] He lost no time in connecting with the agricultural industry and his later research included investigations of a method of purifying muddy waters, of bitter pit and of *Duboisia hopwoodii*, the last two of which were to occupy later researchers in the Department of Physiology as well as Biochemistry. His major work, however, was directed towards the chemistry and physiology of milk. Two of his papers were published in 1914, co-authored with Lilias Jackson, but in his obituary tribute in *Speculum* Osborne noted that the bulk of Rothera's research remained unpublished.[18]

A letter to the University Council before he left for a few months' leave in Europe makes clear the importance and range of his work:

> In a short time I shall be availing myself of the leave of absence granted me but before leaving I wish to express my very sincere appreciation of the accommodation provided for Biochemistry at the beginning of this year.
>
> The new laboratories have proved most useful & have greatly facilitated the work. Since they were ready for use they have been available for research work on Milk, Malting Barley, & the condition of Bitter Pit in Apples.
>
> Agricultural Biochemistry has occupied the main laboratory 6 hours per week throughout the year.
>
> The Biochemistry course for Honours students in Medicine has meant another 3 hours a week during the second term and the Science students taking Physiology have been continuously in the laboratory during the first term and a half.
>
> During the May vacation, a class in Dairy Chemistry was held and ten butter factory managers attended each afternoon for a fortnight.
>
> The laboratory has therefore been in full use, & has I believe fully justified the Government grant.[19]

Rothera's relatively small number of publications may well have been due to his preoccupation with other pursuits. As well as inspiring and overseeing the raising of funds for the extension of the biochemistry laboratory, responding to requests for research and advice from the agricultural industry and acting as professor during Osborne's absence overseas in 1911, Rothera was active in the establishment of the university's Appointments Board. This was formally set up on 3 August 1914 but, because of the outbreak of World War I the week before, did not begin operation until 1920. By this time Arthur Rothera had been dead for five years. He had enlisted in the Australian Army Medical Corps on 11 September 1915, mentioning in his application for a commission that he had three years' experience with the Cambridge University Rifles. He was placed in charge of an isolation hospital at Ascot Vale, which treated convalescent men from the Alfred Hospital. There he contracted influenza and then pneumonia, of which he died on 3 October 1915.[20] A council meeting the following day was cancelled to allow members to attend his funeral. This was as impressive an event as the funeral of Macadam. The cortege travelled from the St Kilda Road base hospital to Coburg Cemetery. The coffin, draped with the Union Jack and resting on a gun carriage pulled by six black horses, was escorted by members of the Australian Expeditionary Force drawn from the Royal Australian Field Artillery and followed by men from the Australian Army Medical Corps, various army officers including the chief medical officer, George Cuscaden, as well as the Chancellor of the university, the Vice-Chancellor, the Dean of Medicine and a number of students.

> Upon reaching the confines of the city the escort and dismounted men, as well as the students, boarded a special tram-car in waiting, and in that manner proceeded to the neighbourhood of the Coburg Cemetery, where they awaited the arrival of the remainder of the cortege.
>
> The service at the graveside was performed by the Rev. D.M. Deasey, Church of England chaplain. The salute was fired and the 'Last Post' sounded over the open grave.[21]

Mrs Rothera, who, as well as having two babies during her time in Australia, corrected examination papers in French for the university, left Melbourne with her two young children in November 1915, returning only briefly and unaccompanied to witness the unveiling of a memorial plaque to her late husband on 18 October 1922; she had remarried in

1920. The plaque was erected in the old Chemistry Building. When this was demolished in 1938, it was placed in the Microbiology Building, later demolished in its turn. After some time in storage the plaque was re-erected where it now stands in the department's foyer in the Medical Building. The attendees of that event, who were addressed by the Dean of Medicine, Professor Sir Harry Allen, included the Chancellor, who also spoke, as well as the Vice-Chancellor and W.J. Young, who was acting Professor of Physiology during Osborne's absence overseas.

Plaque erected in memory of Arthur Rothera, now in the Medical Building

This Tablet is Erected
IN MEMORY OF
ARTHUR CECIL HAMEL ROTHERA
M.A., D.Sc., M.R.C.S., L.R.C.P.
FIRST LECTURER IN BIOCHEMISTRY
IN THE UNIVERSITY
1906–1915.
ENLISTED FOR ACTIVE SERVICE 1915.
DIED 1915.

Despite the considerable attention paid to Rothera's work during his lifetime and immediately after his premature death, his contribution appears to have faded even from the memory of his colleagues in just a few decades. Writing in April 1958 to congratulate a later Professor of Biochemistry on the 'splendid occasion' of the opening, after much delay, of the first stage of the Russell Grimwade School of Biochemistry, one of them commented:

> Ivan and I can look back such a long way to the time when Dr Osborne gave the first lectures in Biochemistry in dark and cramped quarters in the Physiology School and even further to the time when Dr J.F. Wilkinson lectured to the Medical Students to give them some idea that the chemistry of the living cell had some bearing on their clinical work.[22]

For a brief period, Arthur Rothera's position was occupied by the writer of that letter, a young woman who, in a different era, might have proceeded to a stellar career of her own in biochemistry. Lilias Charlotte Jackson (1887–1972) had already, as has been noted, published two papers with Rothera and another with W.A. Osborne in 1914.[23] She had taken her BSc in 1911 and her MSc in 1912, when she also won a Government Research Scholarship of £125 for work on the biochemistry of fish from the nutritional standpoint. As the 1914 winner of the University Scholarship in Physiology (which she held until 1918), she was employed as a demonstrator until August 1915, when she was, at Rothera's suggestion, appointed acting lecturer in his place. He wrote of her in glowing, if slightly ungrammatical, terms:

Lilias Maxwell, nee Jackson, lecturer in biochemistry on her graduation in 1912

> I am able to suggest a substitute whom I honestly believe will carry on the work most efficiently … Miss L.C. Jackson MSc (at present acting as Demonstrator in the Physiology Department & a most successful private coach) will very efficiently take over the work I shall have to leave. It also happens that for the year 1916, owing to a change in the curriculum for the Agriculture Degree, there will be no Agricultural Students doing what are at present 3rd year subjects.[24]

The same year – the first in which women were admitted – Lilias Jackson was elected to the Physiological Society, London. In 1919 she married a fellow member of the Department of Physiology, also admitted to the society in 1915, who was to have a far longer association with the Department of Biochemistry: Leslie Algernon Ivan Maxwell.

As was customary, Lilias Jackson resigned her position when she married. Mrs Ivan Maxwell enjoyed a long career of philanthropy, cultural patronage and motherhood, raising money for University Women's College and presiding over the Catalysts, the Lyceum Club and the

Ivan Maxwell, teacher of clinical biochemistry, 1920–56

University Women's Patriotic Fund, as well as acting, as Victor Trikojus noted, as a 'gracious hostess in the delightful old family house 'Narveno' in Toorak, or at the lovely beachside house at McCrae on Port Phillip Bay'.[25]

Ivan Maxwell (1890–1964) took his first degree in agricultural science in 1914 and his MSc in 1915, being elected to the Physiological Society, London in the same year as his future wife. He had attended University High School and studied agriculture at the University of Melbourne. During his year of practical work at Dookie College, he was awarded the Farrer Gold Medal for an essay on the breeding of wheat. He graduated in medicine in 1918 and was awarded his MD in 1921. His *Clinical Biochemistry* was a standard text, with seven editions published between 1925 and 1947 and a fully revised edition in 1956.[26] Maxwell was appointed to a part-time lectureship in clinical biochemistry in 1920 while retaining a private practice and making regular visits to 'Tara Park', the family's wheat and sheep property in Goombargana, near Brocklesby, NSW. He was a pioneer in the study of allergy, an affiliated fellow of the American Academy of Allergy, a foundation member and president of the Australian Society of Allergists and visiting specialist in allergy to the Repatriation General Hospital, Heidelberg. A plaque honouring his contribution to the department hangs with that of Arthur Rothera in the Medical Building and his portrait by Ernest Buckmaster also hangs in the department.

The Maxwells were both notable philanthropists, and Ivan Maxwell was president of the Friends of the Victorian Symphony Orchestra. He was also an enthusiastic member of both the Wallaby Club and the Boobooks. His association with the department, like that of its foundation professor, spanned the period from when biochemistry was part of the Department of Physiology to its establishment in its own right. Both of the Maxwells' daughters pursued notable careers in the arts. Their son, William Murray Ivan Maxwell (1924–2010), graduated in medicine from Melbourne in 1947, was vice-president of the Victorian Branch of the Australian Medical Association in 1975, chairman of the Federal Assembly of the association from 1977 to 1979 and president of the Thoracic Society of Australia and New Zealand from 1983 to 1985. The largest laboratory of the since-demolished Russell Grimwade School of Biochemistry Building was named the Maxwell Laboratory in 1958. The new Maxwell Laboratory is on level 4 of the Medical Building, where the Buckmaster portrait may be seen.

The Dr J.F. Wilkinson mentioned by Lilias Maxwell in her letter to Victor Trikojus preceded Arthur Rothera in the Department of Physiology and continued to give classes after Rothera's death. John Francis Wilkinson (1864–1935) was appointed honorary demonstrator and lecturer in clinical biochemistry in 1897. He took postgraduate classes from 1903 and was also one of the first clinical biochemists in Melbourne to establish a private laboratory. He was a member of the University Council from 1929 to 1934. Unlike many of his colleagues, Wilkinson is not commemorated in a street name in Canberra; Wilkinson's Lookout at Mount Buffalo, Victoria is, however, named in his honour. Richard Travers gives a succinct account of his principal contributions to clinical biochemistry:

His particular interests were the fractional test meal and the opaque bismuth meal X-ray examination which he introduced with H.M. Hewlett. He was an early advocate of massive doses of iron for anaemia, and popularized the dietary treatment of peptic ulcer and diabetes. His *Synopsis of Lectures Delivered … at the Walter and Eliza Hall Institute, Melbourne Hospital* (1923) was an important contribution to Australian gastroenterology. After seeing insulin used clinically in Canada in September 1922, Wilkinson persuaded the Insulin Committee of the University of Toronto that it should be manufactured under licence by the Commonwealth Serum Laboratories in Melbourne. He demonstrated his initial results to a meeting of the Victorian branch of the British Medical Association in June 1923 and to the Australasian Medical Congress in Melbourne in November.[27]

W.A. Osborne rated the recruitment of William Young to his department as 'the greatest service I have rendered to the Melbourne University'. He came to the university with a considerable body of scientific work behind him, in England and Queensland. William John Young (1878–1942) took his MSc from the university of his native Manchester and his DSc from the University of London. From 1902 until 1912, he worked in the Biochemical Department of the Lister Institute, London, with Arthur Harden (1865–1940). Their discoveries, described in one tribute as epoch-making, 'elucidated the mechanism by which nature attacks and breaks down carbohydrates; by this they made a contribution of inestimable value to biochemistry, for carbohydrates are the fuel of life'.[28] The most significant discovery was that when sugar is united with oxidised phosphorous, it becomes labile and vulnerable.

In 1912 Young was recruited to Townsville to the newly established Australian Institute for Tropical Medicine, now the Australian Institute of Tropical Health and Medicine of James Cook University. W.A. Osborne later described Townsville in terms that will strike a modern reader as racist, as 'a pure white community without coolies or a native population which could act as a reservoir of disease', further stating that 'Nowhere else in the world could the problem be investigated of the action of tropical climates on the white races uncomplicated by tropical diseases'.[29] There, Young worked with Anton Breinl (1840–1944) with whom he published 'Tropical Australia and Its Settlement'.[30] Osborne met Young in 1918 and, evidently anticipating that the anti-German feeling alluded to by Breinl's biographer would make it difficult for their highly successful research collaboration to continue, he asked Young whether he would be prepared to come to Melbourne if a position could be arranged.[31]

Young's investigations were practical and applied as well as theoretical. He worked on the ripening and storage of food, with research workers investigating with him the ripening of bananas, citrus preservation and the cold storage and transport of meat and fish. Young took a prominent place in many professional associations and was sedulous in his attendance at their meetings. Elected to membership of the Victorian Institute of Refrigeration in 1927, he was its president from 1941 until his death the following year. He chaired the Council for Scientific and Industrial Research (CSIR) committee on citrus preservation from 1928 to 1934, was a member of the Nutrition Committee of the Medical Research Council, a member of the council of the Victorian branch of the Australian Chemical Society for ten years, and its president from 1937 to 1939.

His hobbies were gardening, carpentry and walking. He was a sociable man, a keen Rotarian, and, like Osborne before him and Trikojus after him, was a member of the Wallaby Club, which he joined in 1921. He was also very fond of Gilbert and Sullivan operas. Young was elected president of the Wallaby Club in 1925, four years after he joined it, a tribute to a man described with great affection by the colleagues who noted:

William John Young, first Professor of Biochemistry

He believed that a thorough study of the exact sciences should be included in the training of a biochemist, and always emphasised this when advising students. He loved his classes, and always considered difficulties from the student's point of view. This fact, combined with his interest, his quiet sense of humour, and willingness to assist won him the affection of all those who came in contact with him.[32]

William and Janet Young's daughter, Sylvia, took her degree in medicine from the University of Melbourne and spent several years with the Medical Colonial Service, working in Palestine from 1936 to 1945 and afterwards in Sierra Leone. Her principal interest was in women's and infant health, and the fact that she was a woman was naturally of particular benefit to Arab women precluded from being examined by a male doctor.[33]

W.J. Young's early death in 1942 left the Department of Biochemistry without a professor at a time when it was highly involved in research of considerable importance to Australia's engagement in World War II.

Sir Robert Menzies and Victor Trikojus at the unveiling of the
portrait by Louis Kahan, 1968

CHAPTER 2

The War Years

WORLD WAR II was well underway when the department was devastated by the unexpected death of the professor who had led teaching and research in biochemistry at the University of Melbourne for almost twenty years before its establishment, and almost half a decade since. His passing left the department with a heavy teaching load, existing research projects to be completed and new tasks to be undertaken without its highly competent, respected and hardworking head. The decision to appoint a local man to replace William John Young may have been influenced by considerations of economy and the speed with which he might be able to take up his appointment, as well as by his record of achievement, but Victor Martin Trikojus was to lead the department with distinction for a quarter of a century and enhance its international reputation.

Biochemistry is central to almost all areas of scientific work. Although the Department of Biochemistry and Molecular Biology at Melbourne is administratively located within the structure of the Faculty of Medicine, Dentistry and Health Science, its undergraduate students come from the faculties of Science, Agricultural Science and Veterinary Science and also include prospective doctors and dentists. Postgraduate research in biochemistry is now pursued less frequently by medical graduates than by chemists, agricultural scientists, botanists and zoologists. This was reflected in the establishment in 2005 of the University's Bio21 Molecular Science and Biotechnology Institute, which is a multidisciplinary research centre specialising in medical, agricultural and environmental biotechnology. It is not, however, a new development. Although both W.A. Osborne and Arthur Rothera held medical qualifications, they both showed a decided preference for science. His biographer in the *Australian Dictionary of Biography* tells us that Osborne's father had pushed him into medicine but he 'escaped into science'.[1] Rothera told Gowland Hopkins before he took his medical degree that he 'would look to biochemistry for a career'.[2] It was merely because of the war that he died practising as a physician.

During World War I, although the biochemists within the Physiology Department did some important work, many highly promising young scientists such as Arthur Rothera and Clunes Mathison, dead before he could take up the post of director of the Walter and Eliza Hall Institute (WEHI), are remembered principally for their untimely passing. In World War II, working in 'reserved' occupations, scientists were able to conduct significant investigations with practical applications to the war effort.

The work required of the university during World War II amply demonstrated the intimate links between biochemistry and other scientific and medical disciplines. What was also notable during this period was the very high degree of interdepartmental and interdisciplinary cooperation and collaboration, as exemplified in the case of the Optical Munitions Panel which incorporated personnel mainly from Physics, Chemistry and Botany, with input from Biochemistry and Physiology. As the university's *Annual Report 1939–1946* noted, a product originally developed to preserve blood plasma was found to have another application:

When the Tropic Proofing Committee of the Optical Munitions Panel was seeking a suitable fungus to protect optical instruments from moulds, Professor Trikojus suggested the use of sodium-ethyl-mercuri-salicylate, a compound then being synthesized for the Forces in the Biochemistry Department for use in preserving blood plasma.

When its efficacy had been proved, quantities of this material were made for distribution to the Services in Australia and New Guinea, and the Biochemistry Department remained the sole source of supply in Australia of this valuable material.[3]

Merthiosal (originally named Merthiolate) was effective in preventing fungal growth on glass without causing corrosion of the metals in the instruments. Humphreys tells us that:

By the beginning of 1944 the US airforce reported that none of the 350 cameras being treated with Merthiosal became infected at a time when fifty untreated cameras were returned for repair. In 1944 all types of optical instruments manufactured or assembled for use by Australian army, navy and airforce were treated with Merthiosal to great effect.[4]

The department's aid was also required in the medical testing of men enlisting in the army:

Early in the war glucose tolerance tests were carried out on recruits for the A.I.F. who showed glycosuria. Vitamin estimations, mainly of vitamins B and C, were undertaken on foodstuffs and concentrates likely to be of use by the Armed forces. The department also analysed certain solutions and prepared chemicals required by military hospitals.[5]

There was also collaboration between universities, notably with the new professor's alma mater:

An extensive drug programme was commenced. In association with the Drug Group of the University of Sydney, and assisted by the Chemistry School at Melbourne, methods for the production of adrenalin, eserine substitutes and benzidrene were investigated. Other drugs studied ranged from those with action against the typhus organism to those with an anti-histamine effect likely to prove of value in defence aspects of chemical warfare.[6]

This new professor, who joined the university's war effort to such good effect, had been through a difficult time since the outbreak of hostilities, despite being Australian born. Victor Martin Trikojus (1902–1985) took his BSc in 1925 from the University of Sydney, his PhD in 1934 from Oxford and his DSc in 1956. Readers wanting a fuller account of his life outside the department should consult *Trikojus: a Scientist for Interesting Times* by L.R. Humphreys and the obituary tribute by J.W. Legge and F. Gibson.[7]

Victor Martin Trikojus, Professor of Biochemistry, 1943–68

Trikojus was born in Sydney, the child of an Australian mother and a Prussian father who died when Victor, the oldest of three children, was only nine years old. He went to Sydney Technical High School and studied organic chemistry at the University of Sydney. After a brilliant under-graduate career followed by study in the United Kingdom and Germany, Trikojus was appointed lecturer at Sydney University in the Department of Organic Chemistry from 1928 to 1932 and in medical organic chemistry in the Department of Medicine from 1932 to 1943. In 1925 he had been awarded an 1851 Exhibition Science Research Scholarship and spent 1926 to 1928 at Oxford and the Laboratorium des Staates in Munich.

His professional and familial links to Germany, some possibly unwise remarks about the economic efficiency of the Nazi regime in its early days and a degree of animus on the part of other scientists are probably all to blame for the decision of the Australian authorities to arrest and detain a man in whom the Australian Government was later to repose an extraordinary degree of trust. He had already done highly valued work for the Drugs Subcommittee of the Australian Association of Scientific Workers, which provided advice to the Medical Equipment Control Committee of the Department of the Army. This did not prevent his arrest and internment in the State Reformatory from 12 February to 18 April 1941 as detainee number PWN 1445. Despite a clear statement on the Australian Military Forces documents that he was born in Sydney, the papers were firmly stamped with the heading 'GERMAN'. This episode might have been expected to take its rightful place in the history of wartime outrages such as the treatment of the *Dunera* refugees and the university's Japanese instructor, Moshi Inagaki, but it had a long life, resurfacing in an article published in 1999 which implied that Trikojus had Nazi sympathies.[8]

The allegation was swiftly rebutted by his former colleagues Francis Hird and Max Marginson, who noted that his central role in the production of sulphaguanidine, a dysentery-treating drug that saved the lives of many Australian soldiers in the Pacific, caused many survivors to regard him as a hero of Kokoda.[9] This view was supported by Sir Alan Newton, chair of the Medical Equipment Control Committee. It is notable that the process for producing sulphaguanidine and other drugs was made available to industry – in the case of the dysentery treatment, to Monsanto – without requiring the payment of royalties.

A major problem of the period was the pre-eminence of overseas chemical laboratories and manufacturers and the difficulty of ensuring supplies of chemicals for drug manufacture in Australia. Trikojus, before coming to Melbourne, had already brought to successful conclusion investigations to prevent scurvy through the large-scale production of ascorbic acid based on sorbitol, and others concerning an arsenical drug to treat syphilis, which was marketed as Carbasone. The man who agreed over lunch at the Melbourne Club with Alan Newton and Hugh Trumble to apply for the Chair in Biochemistry left suddenly vacant by W.J. Young was therefore already known throughout the biochemistry departments of Australia and indeed of much of the world.

The Melbourne Department of Biochemistry of the time was tiny. In 1940, apart from the professor, it counted one full-time lecturer, Kathleen Law; one part-time lecturer, Ivan Maxwell; a full-time demonstrator, Jean Millis; and some part-time demonstrators. The secretary, Meriel Wilmot, was shared with the Department of Physiology, which also shared part of the accommodation. In 1943, the newly appointed Victor Trikojus was supported full-time only by Kathleen Law and Muriel Crabtree as demonstrator, with part-time assistance from Ivan Maxwell, Jean Millis and Cecil Hearman, who lectured in preventive dentistry and dietetics. Nonetheless, Trikojus would become a dominant force in Australian biochemistry of the time. Moreover, his friendly relationship with Sir Russell Grimwade, and the respect the two men had for each other, would transform the department.

Remembered by Max Marginson as one of the last great god-professors, Trikojus joined a growing group of young heads of department in the Medical School. R.D. Wright had been appointed to the Chair of Physiology in 1939 and Sydney Sunderland to that of Anatomy the following year. Sydney Rubbo would join them in Bacteriology in 1945 and Edgar King took up the Chair of Pathology in 1951. Francis Hird, who joined the department in 1951 and was professor from 1964 to 1985, commented on Trikojus's determination to foster research and his skill in raising the funds required, while displaying little of the bonhomie evinced by some other senior university staff:

> Concerning the Department, it was a tight ship – clean, no nails in the walls, all notices authorised, central control, everyone knew where they stood. Nobody went to him without a real demand. No possibility of independent empire building. The technical staff respected him and so did we all for that matter. This went for the central administration and the Maintenance Department also …
>
> It was a smaller field in his day, but there were very few departments that one could visit abroad where he was not known. As he passed through on overseas visits, he always left a strong image behind. At the University level, Trikojus was very influential …
>
> What Trikojus provided for his staff was patronage. There was never any major decision for me to take or any that I was encouraged

or permitted to take. On the other hand, no person of quality in his department ever lacked the facilities and equipment with which to do good work – his own work – and that was a lot.[10]

It is strange, in the knowledge that the end of 1944 would see Sir Russell Grimwade's offer of £50000 for a new building for Biochemistry, to read in the Trikojus archive of the desperate borrowing and lending of equipment necessitated at the same time by the war. On 2 June 1943, the secretary of the Commonwealth Scientific Liaison Bureau wrote to Trikojus on behalf of the Department of Commerce and Agriculture, asking for the loan of apparatus either for the duration of the war or at least until alternative supply could be arranged. On 7 June 1943 he was offered (subject to the Vice-Chancellor's approval) 'an analytical balance – air damped – (to 1 mg) Sartorius; a small electric muffle furnace, 6½ long × 3" wide × 2" high (requiring repair); three Platinum dishes (c. 6 centimetres in diameter); a Zeiss Photoelectric colorimeter and a Klett Biocolorimeter'.[11]

Trikojus requested return of the dishes and the Sartorius air-damped balance in January 1944 because extensions were being made to the department's laboratory accommodation and stocks of equipment needed to be divided between locations. It was only in March that the Department of Commerce and Agriculture responded, promising to return the equipment forthwith, although adding: 'We had expected daily to learn that we could obtain a balance of similar type, but have only just been advised that the source on which we had counted will not yield us anything'.

A Zeiss refractometer proved equally troublesome. In March 1944 Gerald Lightfoot (1877–1966), the secretary of CSIR, acknowledged a letter requesting its return but, pointing out that it was still being used by the Division of Food Preservation and Transport for urgent work in connection with defence foodstuffs, asked for the loan to be extended: 'Some delay has occurred in obtaining an instrument of this type through our customary channels'. Trikojus agreed, undertaking to try borrowing one for use in Biochemistry from another department. This proved unsuccessful, however, and in April 1945 the device was regretfully returned.[12]

Even before Trikojus arrived in Melbourne, the Medical Faculty minutes reveal a department under considerable stress. Kathleen Law was appointed acting senior lecturer and de facto Head of Department in June 1942 after the death of Young, with an increment of £150 in 1943 'until the new Professor assumes office'. It was therefore she who had to reply to a December 1942 memorandum from the registrar, warning that:

> It is impossible to make the usual arrangements for scrubbing this long vacation as so few women are available.
>
> If heads of departments could arrange for their own cleaners to scrub studies it might be possible for the offices to find enough casual hands to do the theatres, passages and class rooms.[13]

Given the university's documented reluctance to pay its female cleaners more than a pittance, it is perhaps a good thing that Law was able to reply that only the agriculture laboratory required scrubbing, and even there it could be dispensed with.[14]

In a department in which minor accidents involving spills were not uncommon, however, protective clothing was imperative. But this too was in short supply. On 5 August 1942, Law informed the registrar:

> In this Department only the three technical assistants have laboratory overalls issued to them. The number averages one per year per person. In one case (Miss Richards) these are white cotton or drill material. In the two other cases (Mr Egan and assistant) coats of holland or similar material are preferable; the type varies. The University pays for these and they are the property of the Department. Such overalls are necessary for the protection of ordinary clothing.
>
> Two men's laboratory overalls are required before next term.
>
> The teaching and research staff provide their own overalls, and these are their personal property. They require them for protection of ordinary clothing, and should be classed under Section D.[15]

The day after the University Council had formally registered its thanks to the staff of the department for carrying the load since the death of W.J. Young, Kathleen Law, in her capacity as house warden, listed the air-raid precautions in force:

Personnel:

Miss K. Law, Janet Clarke Hall (F. 2208)

Miss Millis, Drake Street Brighton (X.A. 1208) (attested, not enrolled)

First aid:

Miss M. Crabtree, University Women's College (F. 2539)

Miss J. Dickinson, 50 Canterbury Road Camberwell (W.F. 1733)

Equipment:

3 Stirrup pumps

6 Water buckets

5 Sand buckets

2 Arm bands

3 Shovels

3 Shields

3 Hoes[16]

Matters did not improve with the arrival of the new professor. Finding space to accommodate any laboratory equipment and the people to make use of it, including large classes of undergraduates, was extraordinarily hard and made even harder by the requirements of the Ministry of Munitions, which was conducting research into physiological aspects of chemical warfare in rooms that Trikojus had hoped to occupy. There were some 360 students, and the existing laboratories (some shared with Physiology and also in some cases with animals) were decrepit, with bits of plaster falling from the ceilings and sinks continually blocking, with consequent flooding of other areas. Although he fought valiantly and successfully for better classroom accommodation and was himself a fluent and cogent lecturer, Trikojus was above all determined to maintain, if not improve, the research reputation of his department. To that end he insisted on the recruitment of new senior staff and welcomed a number of outside stars to temporary positions. Apart from Ivan Maxwell, the lecturer in clinical biochemistry, all the wartime staff, until W.A. Rawlinson was appointed in 1944, were women.

Trikojus's own investigations after the war concentrated on thyroid research. Legge and Gibson have speculated that this was motivated in part by a desire to contribute to the reduction of human suffering, in part by a recognition that funding for research was most likely to come from the National Health and Medical Research Council (NHMRC), and in part because a field which was both of medical importance and scientifically challenging was likely to attract the attention of the student population. They quote his 1944 Syme Lecture: 'Biologist, clinician, microanalyst, chemist and, latterly, the atomic physicists have all become attracted by the mysteries of iodine metabolism, and it is probable that only through such combinations of specialised interests will the aetiologies of thyroid dysfunction become clarified'.[17]

During his first ten years at Melbourne, Trikojus investigated approaches to the isolation, separation, identification and quantitative analysis of thyroxine, its precursors and products. It was not until 1946 that he was able to publish the results of his work. Four papers appeared that year. One of the most cited, co-authored with Muriel G. Crabtree, was on the thyrotrophic hormone, thyroxine and ascorbic acid in relationship to the liver glycogen of guinea pigs.[18] A further four papers followed in 1948. In Trikojus's second decade at Melbourne, he directed his attention to the nature of the protease that split thyroxine and the iodotyrosines from the parent thyroglobulin.

As is so often the case with senior university staff, by the time he reached his third decade at Melbourne, Trikojus's time was increasingly taken up with administration, both within his own department and in the wider university community. During the war he had chaired the Drugs Subcommittee of the Australian Association of Scientific Workers and been director of research of the Medical Equipment Control Committee. From 1946 to 1957 he was a member of the NHMRC Medical Research Advisory Committee and a member then chairman of the Natural Sciences Section of Unesco. In 1954 he was an inaugural member of the Australian Academy of Science. He served from 1962 to 1967 on the council of CSIRO, was the first professorial Dean of Graduate Studies in 1963 and 1964, and a foundation member of the Australian Research Grants Committee. He chaired the Professorial Board in 1967. He was a foundation member and one of the first honorary members of the Australian Biochemical Society.

Victor Trikojus was also an active member of the Melbourne University Chemical Society, addressing its meetings four times between 1944 and 1952. His first address was given to an Annual Combined Meeting of MUCS, the Society of Chemical Industry of Victoria and Royal Australian Chemical Institute, in March 1944: he spoke on the chemistry and physiology of vitamin C. The following year he addressed the September meeting on photosensitising pigments with particular reference to hypericin and related substances.[19] His staff also addressed the society, with W.A. Rawlinson (a Melbourne biochemistry graduate recruited from a position at the WEHI in September 1944) speaking on the application of magnetic measurements to chemical problems – magnetochemistry – in October 1943. Rawlinson was followed in August 1946 by J.W.H. Lugg, who addressed the audience on some fundamental problems in the amino acid analysis of proteins.

Retirement in 1968 by no means signalled the end of university work for Victor Trikojus, although after such a long career in sole charge of a department he could very much regard as his bailiwick, it was desirable for him to find appropriate work accommodation in another. Appointed emeritus professor, he was made an honorary professor in the School of Chemistry, where, with D.W. Cameron and W.H. Sawyer (who would later follow him as Head of the Department of Biochemistry), he did several days' work a week at his bench and published his last paper in 1977.[20]

The close links between Biochemistry and other departments such as Agriculture, Botany and Chemistry have already been mentioned: another especially productive link was maintained with what was later to become the Division of Protein Chemistry of CSIRO. The Biochemistry Unit of the CSIR Division of Applied Chemistry was headed by a graduate of the Biochemistry Department. Francis Gordon Lennox (1912–1998) was one of the external luminaries who lectured there part-time during the 1950s. Lennox, who won the Grimwade Prize for Chemical Research in 1943, had taken his DSc in 1941 from the University of Melbourne. He spent over thirty years with CSIRO, the last fourteen as chief of the Division of Protein Chemistry.[21]

Francis Gordon Lennox, Deputy Chancellor of Monash University, 1968–73

In 1946 Trikojus appointed a secretary who was to remain with the department for thirty-six years, rising from that position to senior administrative assistant. Nancy Ethel Hosking (1922–) replaced Marion Goodwin, who had initially taken over the part-time services of the Physiology Department's Meriel Wilmot from July 1943 to early 1946. The position may not initially (while the total staff remained small) have been onerous, since Meriel Wilmot Wright commented in 2013 that she could not recall typing anything for either W.J. Young or Trikojus.[22] The advertisement for the position fell a long way short of describing what evolved into that of confidential secretary, part-time librarian and personal assistant, stating:

> The successful applicant to provide evidence of being expert in short-hand and typing, and preference will be given to an applicant with experience in dealing with scientific material. If desired, the appointee will be permitted to undertake a University course on a part-time basis. Salary range £234–£250.[23]

Nancy Hosking was very highly esteemed. Humphreys tells us that she was 'universally admired for her supportive competence, diligence, sensitivity and tact' and that she was 'affectionate and consistently supportive over a long period and her commitment extended into Trikojus's retirement years'.[24]

The University of Melbourne career of Archibald Macdonald Gallacher (1925–2010) was almost contemporaneous with that of Victor Trikojus himself. Gallacher came to Biochemistry in 1946 and left to take the position of laboratory manager of the Chemistry Department at La Trobe University in 1966. The fact that the 21-year-old Gallacher was hired as a junior technical assistant obscured an employment history which was more varied than one might have expected. From 1940 to 1943 he was employed at the WEHI as a laboratory technician under Henry Holden, the Head of Biochemistry, where he worked on the development of suitable methods for the collection and analysis of atmospheric samples from military vehicles and aircraft, while at the same time taking some subjects of the Diploma of Applied Chemistry. He enlisted in the RAAF in October 1943 and served as a radio operator/emergency air gunner with the rank of flight sergeant until March 1946. He was appointed to the Department of Biochemistry three weeks later.

Initially appointed to assist J.W.H. Lugg, Gallacher climbed steadily up the career ladder from technical assistant to laboratory manager, although Trikojus's recommendations for his promotion were all too often knocked back. In 1959 the professor made some of the problems facing Archibald Gallacher clear, writing:

> As you are aware, the Department of Biochemistry handles more students, except for those in first year, than any other department in the technical Faculties, including Physics and Chemistry; moreover, it has extensive research interests. In these various activities I rely upon Mr Gallacher to perform the duties of Laboratory Manager, a task which has not been made easier by the division of the Department between two buildings. Apart from this, he has, over the past four years, relieved me of much of the detail of the planning for the Russell Grimwade School of Biochemistry and is at present very actively concerned with the final stages of the preparation for the calling of tenders for the upper floors. Mr Gallacher has also the general supervision of some twenty-three members of the technical and cleaning staff.
>
> Although it is in direct relationship to my own responsibilities and those of the Department that I rate Gallacher so highly, I do know that other senior members of the Department are impressed by the efficient manner in which his services are given. I might add that, on the work in connection with the new School, he has often spent every night of the week and weekends in going over details of building plans.[25]

The varied nature of the duties required of members of what were called the general staff of the time can be seen in a letter from Trikojus in America to Gallacher in 1952, when the senior technician might have been expected to be very busy preparing for the start of a new academic year on 31 March:

> This is just a brief note to ask whether you would kindly meet Mrs Trikojus and the children when they arrive in Melbourne, which according to the ship's schedule is March 16 (the P. & O. Strathaird).

I am sure you will be a great help to them as you can imagine it is bad enough for an adult without children around also. Would you also kindly inquire from Professor Dunkin of the Department of Mining what the position is in regard to the tenants who are in our house?[26]

Archibald Gallacher himself pointed out in his application for a position at La Trobe University that by 1966 he administered fifteen different accounts supporting three professors, fifteen other academic staff, thirty graduate research students, twenty-seven technical staff and six cleaners. He acknowledged the help of a secretary. Gallacher retired from La Trobe, where he supervised twenty staff, in 1986, and in 1987 he was awarded the La Trobe University Distinguished Service Medal for his outstanding service.

As the war drew to a close, the Department of Biochemistry could look forward with optimism to increased support for its endeavours. A letter dated 6 November 1944 from Russell Grimwade, who had been a member of the University Council since 1935 and Deputy Chancellor from 1941 to 1943, offered £50 000 to the university for a biochemistry building. On 2 July 1945, the council recorded an offer from Nicholas Pty Ltd of £20 000 for the erection of a School of Nutrition, contingent on the government's furnishing and equipping the school so that it could function in a practical and efficient manner; £5000 had been paid as the first of four projected annual instalments. Postwar shortages, university prevarication over the placement and design of the building eventually known as the Russell Grimwade School of Biochemistry, and the failure of government support to materialise as fast as had been hoped meant that the new building, incorporating the Nicholas Nutrition Laboratories, would not be built in Grimwade's lifetime.

The building was desperately needed. By 1950 the department's activities had been spread over four separate buildings and on 30 October Trikojus wrote to a staff member travelling overseas:

You must have heard that the Department had a fire at last, but succeeded in burning out only one laboratory so that the new school has no added incentive. Still, surveying the mess and damage, which has taken 4 months to repair, I decided I did not want to accelerate building that way. Perhaps we shall start in earnest next year – at least we have the assurance of double the original funds from the State Government to offset increased costs.[27]

Plans for the building, however, were protracted, complicated by both government and university policy. As late as October 1951, Trikojus was writing to J.S. Turner of the Botany School that he was astonished to learn that the Vice-Chancellor believed that it was likely to be ten years before building started. Professor Turner himself gave some background to this, together with a snappy view of the university's master plan of the time:

> In some ways the building site suggested opposite the Royal Melbourne Hospital is a good one, but I think it has two disadvantages. In the first place it might lead to Biochemistry becoming surrounded in time by large buildings … secondly the construction of a large building on that corner site would more or less commit the University to Lewis's somewhat grandiose scheme of turning the University into a kind of enclosed backyard with high buildings all around.[28]

The first stage of the Russell Grimwade Building was opened in 1958, and further government support was needed as well as another donation of £40 000 from Grimwade's widow before it could be completed in 1963. In all, it cost in the order of £700 000. It would close its doors for demolition in 2007.

Despite the seemingly interminable delay in getting an adequate building, the decades following the war were to be satisfying for the department. Already, in 1944, W.A. Rawlinson had been recruited from the WEHI and at the end of 1945 J.W.H Lugg left his position at CSIR to take up a lectureship in plant biochemistry. They joined a talented cohort of women scientists.

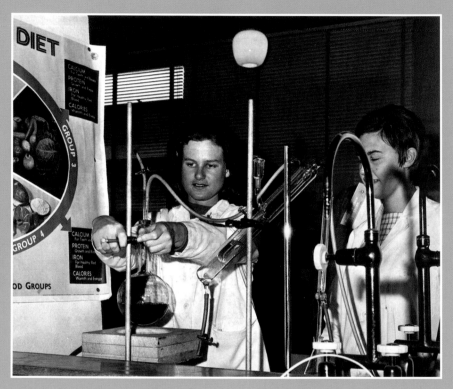

The Nicholas Nutrition Laboratory in the Russell Grimwade Building

CHAPTER 3

Trailblazing Women

Mary-Jane Gething, who went on to become the first woman to head the Department of Biochemistry at Melbourne University, noted in an oral history recorded in 2003 that the department counted among its staff and research students an unusually high number of women. This was certainly true during the World War II years and immediately afterwards, and unlike many of the women who attained mid-level positions during World War I, those who rose to the position of lecturer and above during World War II did not necessarily retire when they married or because a man might have become available to take a job. The photograph of the Department of Biochemistry in 1948 shows a teaching and research staff of twelve people, five of whom are women.

Pre-eminent among the female staff before, during and immediately after the war was Kathleen Alice O'Dell Law (1905–1994), the daughter of Dr Archibald Law, the vicar of St John's Toorak. She was awarded a Final Honour Scholarship for 1928/29 and appointed demonstrator the following year. Granted fifteen months' study leave in 1934, she spent some time in the United States and Europe, including a term at the Lister Institute, at which she co-authored two papers published the following year in *Biochemical Journal*.[1] On her return, like many women of the time, she took on teaching outside the department, working as a tutor in Janet Clarke Hall. In 1938 she was promoted to lecturer, and following the death of W.J. Young in 1942 she was temporarily promoted to senior lecturer and de facto Head of Department, but reverted to the lower classification when Victor Trikojus arrived.

This did not meet with his approval and on 12 May 1947 he wrote to the Staff and Establishments Committee, unsuccessfully arguing for her permanent promotion to senior lecturer:

> It is difficult to express the value of Miss Law's influence in the Department; she undertakes lecturing duties both for advanced students and those in the early years of their courses. Moreover, she contributes very largely to the organization of the practical work of the Department. All these duties she performs most conscientiously and with credit to the University. It is true that Miss Law has not been able to contribute to scientific literature during the past few years, but it can be pointed out that until fairly recently the entire lecturing work of the Department was undertaken by the Professor and Miss Law, who, in addition, took over a large amount of responsibility for the practical work. Since the enlargement of the staff of the Department by the appointment of two Senior Lecturers and an additional Lecturer, Miss Law's formal duties have decreased, and she has found time during the past eighteen months to carry out experimental work of an original nature … Personally, I owe a great debt to Miss Law for her fine co-operation.[2]

Kathleen Law resigned in May 1949, and although she published a paper in 1950 in which she is identified as a member of the department, she had already joined the Biochemistry Department of University College London, where she published papers in 1955 and 1956 on wood-rotting fungi.[3] She does not appear to have published after that time, and returned to Australia in 1970.

Jean Millis Jackson (1913–2012) was the elder sister and mentor of another famous bioscientist, the former chancellor of La Trobe University, Nancy Millis AC, MBE (1922–2012). Jean Millis graduated BSc in 1935 and MSc in 1936. She began her academic career in the Department of Physiology.[4] In March 1935 she was appointed temporary demonstrator in Biochemistry on the recommendation of W.J. Young to help cover his absence during 1936. The same year she was awarded the Dunlop Rubber Company Exhibition for Biochemistry with Bacteriology. Having been appointed demonstrator to science and agricultural science students and giving classes in nutrition in 1937, she became senior demonstrator in clinical biochemistry in 1938. Jean Millis remained with Physiology until 1943, when she transferred to the Department of Biochemistry under Trikojus,

Jean Millis Jackson, lecturer in nutrition and dietetics

with a brief interlude in the United Kingdom on a British scholarship which paid her more than the £250 a year paid by the university. This caused some problems on her return, but the university was finally induced to pay her at the same rate she had been receiving in England.

Kathleen Law published more after her departure from Melbourne than she had found time to do earlier. Jean Millis Jackson made clear in an interview in 2010 how severely the enormous teaching load carried by female staff could impact on their capacity to undertake research and publish results.[5] Jackson herself taught six days a week, supplementing her university salary with private tutoring. Even that would not have been enough to permit her to live outside the parental home and certainly left very little time for research. In 1949 applications were invited to teach nutrition to fifth-year students in social medicine and hygiene at the University of Malaya in Singapore, and in his letter recommending her for the position Trikojus showed a clear appreciation of this fact:

> Miss Millis has organized instruction with enthusiasm and has given the students first-class training. She has been a prominent member of the Dietetic Association and has interested herself in other ways with nutritional matters in the State. In spite of fairly heavy teaching

commitments she has managed to find time for original investigation. One recent study has been concerned with the effects of prolonged feeding of heated fats to the laboratory rats. No detrimental effects were observed after a short period of fat administration but after 18 months the groups of rats on the heated fat diet showed significantly higher mortality rates and pathological changes, particularly in the kidneys. An examination of the post-mortem material has not yet been completed.

In addition to her usual duties, Miss Millis has this year given the series of some 40 lectures in elementary biochemistry to students in the second year of medicine. She has devoted much attention to the preparation of the lecture material and I have learnt that the students have appreciated the course given by her.

Miss Millis is a most conscientious person and is reliable and accurate in her work. She is well read in her subject and interested in the advancement of it.[6]

The opportunity to travel to Singapore had come only after a terrible few months the year before. In 1948 Nancy Millis was in Papua New Guinea to study the agricultural practices of women there when she developed peritonitis and almost died. Jean left Australia to care for her, but her concern for Nancy did not mean that she forgot either her students or the work of the department, especially the marking of examination papers. On 10 November Trikojus wrote to reassure her:

We have made arrangements for both Part I examinations. I am sure the tropics do not provide a suitable atmosphere for such chores. You will probably be back in time to do the Nutrition papers, but I shall send them on if you wish … Regarding the lectures, I hope you will be back in time to give them as I am sure it will be better for the graduates if they hear them from you.[7]

The following week, he offered some practical help:

The continued state of your sister's health must be very disturbing. Would you care to send down a summary of the main events in her illness? You may wish me to discuss it with some of the senior consultants here. I should be very glad to help in this or in any other way that you suggest.[8]

Thanks to streptomycin sent from Melbourne, Nancy Millis recovered and Jean was able to return briefly to the department. At the University of Malaya, from which she took her PhD, her research output soared: she published almost a dozen papers between 1954 and 1958. In 2001 Nancy described her sister's work:

> She was interested in the nutritional intake of children. Many people in Singapore were pretty poorly nourished, and there was interest in assisting to improve the diet of the various ethnic communities: Chinese, Indian and Malay. The Chinese were much easier to help, in that they had a broader tolerance of what foods they would actually eat, but certain groups had religious problems, particularly, with eating such meats as pork. Those could simply not be added to the diet.
>
> My sister had a team of young folk – Chinese or Malays or Indians, as the case might be – who went around into the various shop-houses in Singapore to weigh children in their homes. There's no way that the mothers would go to a clinic as you might go to a health centre here, so the team had to go to them. When I visited my sister on one occasion I went around to see how it was all done. These little shop-houses would be no more than about 10 or 12 feet wide. You'd walk through whatever activity was going on in the shop, right through to the back, upstairs, and off a passageway, there would be small rooms for three or four families, perhaps, living in the rooms. So you might find that a bed for father and mother would have then an underlayer and an overlayer and a side layer as well.[9]

Millis's papers appeared in a variety of journals, notably the *Medical Journal of Malaya*, *Quarterly Review of Pediatrics* and *Annals of Human Genetics*. The last is one of the most cited: published in 1954 and co-authored with You Poh Seng of the University of Malaya Statistical Unit, it deals with the effect of age and parity of the mother on birth weight of the offspring, based on a survey of 18 425 Chinese infants born in Singapore in 1950 and 1951.[10] Later papers are comparative studies between different ethnic groups.

In 1957 Jean Millis's career took an unexpected turn, best described in her letter to Trikojus written on 5 February:

> Just a note to tell you I have made devastating changes in my plans for the future: I married Dr Arthur Jackson – Director of Chemistry, Malaya – on Wednesday 10th Jan. I don't know who is most surprised – my friends or myself. Our engagement lasted about a week & we had told relatively few people about that either. I am afraid the family will be quite rocked.
>
> Jack has been asked to stay on after malayanisation and we will continue to live in Singapore for 2 or perhaps 3 years – but I don't know whether it will be Aust. or Britain after that … I do hope my change of plans has not upset any arrangements you were making. I can assure you I had no such ideas when I saw you or indeed 3 weeks ago. However I can tell you that after nearly a week of married life I approve of it & wonder why I didn't think of it before.[11]

Arthur Jackson had worked as a forensic scientist in Hong Kong before the war and attributed his survival in the notorious Changi prisoner-of-war camp to his knowledge of plants and chemistry. On their return to Australia in 1960 the Jacksons settled in Mount Eliza, where 'Jack' Jackson taught science and mathematics. Jean Jackson returned part-time to the Department of Biochemistry and was promoted to senior lecturer in 1963, but finding the daily trip from Mount Eliza too onerous she resigned in 1966 in favour of occasional contract work with the World Health Organization.

Jean Millis Jackson's interest in nutrition and dietetics had a long ancestry in the Department of Biochemistry. Ethel Elizabeth Osborne (1882–1968), the wife of W.A. Osborne and a distinguished medical practitioner in her own right, was the joint author of their 1910 *Primer of Dietetics*. During an overseas trip in the late 1920s she investigated dietary departments and schools on behalf of St Vincent's Hospital, in which the first dietetics school in Victoria was established.[12]

All of the Osbornes' four children distinguished themselves professionally, but it is their eldest child, Audrey Josephine Cahn (1905–2008), who is of greatest interest to historians of the Department of Biochemistry. She was a true 'university child', having been born in the family's apartment in the West Wing of the Quadrangle previously occupied by Frederick McCoy. Audrey Cahn took her BAgrSc from the University of Melbourne in 1928, the first woman to take the degree. She also won exhibitions in botany and zoology. In 1930 she was awarded the M.A. Bartlett Research Scholarship in Physiology of £100 for work on

Audrey Cahn, lecturer in Nutrition and Dietetics

fat particles of milk, with another £5 awarded for research or special apparatus for her investigations of oil globules of milk. On graduation she turned down what she assumed would have been a desk job with the Department of Agriculture in favour of a position as microbiologist and food analyst with Kraft. In 1929 she married the architect Leslie Cahn in a wedding attended by the Premier of Victoria and virtually every senior officer of the university.[13] The couple's twin daughters were born the following year.[14] The marriage did not last and Cahn was faced with the need to support herself and the children. Having enrolled in the Dietetics Unit inspired by her mother at St Vincent's Hospital, she stayed on as chief dietician. After work as a microbiologist with the Kraft/Walker Milk and Cheese Factory in Drouin, which necessitated a weekly round trip of some 200 kilometres to visit her children, she took various positions as a dietician in Melbourne and Perth and was awarded her Diploma of Dietetics by thesis in 1937.[15]

Cahn enlisted in the Australian Army Medical Corps in 1943, serving until 1946 as chief dietician at the Heidelberg Military Hospital with the rank of major. In 1947 she was appointed lecturer in dietetics and began what was to be a twenty-year career in the university. On Jean Millis's departure, she was given responsibility for both the nutrition and dietetics courses. The three-year Diploma of Dietetics course encompassed various units of the BSc as well as units in food chemistry and preparation, food services and administration, and nutrition. W.A. Osborne had put the first proposal for a course in dietetics and nutrition to the University Council in October 1930, but despite his daughter's best endeavours and the offer from Nicholas Pty Ltd in 1944 of £20 000 for the erection of a School of Nutrition, this was not to materialise for over a decade. It was finally housed on the fourth floor of the Russell Grimwade School of Biochemistry Building.

Cahn had an extraordinary workload, giving all the lectures in nutrition and initially running all the practical classes herself. She was promoted to senior lecturer in nutrition and applied dietetics in 1959. The position of her discipline within the department was somewhat ambiguous as it was regarded as a 'soft' science of overwhelmingly female interest.

Before the construction of the Russell Grimwade Building, the laboratory accommodation was seriously inadequate:

> There was no electrophoresis, only elementary chromatography, very limited spectrophotometry, very limited access to radiolabels and no Sigma catalogue of chemicals. All biochemical reagents had to be prepared in-house and the practical laboratory smelled of gas and had only two power points and very little equipment.[16]

In fact, in 1947 the Sub-Dean of the Science Faculty reported to the registrar that Janet Clarke Hall had kindly offered the use of their special diet kitchen for practical classes.[17] The situation improved when the department moved into the new building, although some members complained of the cooking smells which emanated from the fourth floor.

In 1954 Cahn established a long-term dietary survey of a selected group of metropolitan children, which formed part of the Melbourne Child Growth Study (1954–71), in collaboration with the Department of Anatomy. The data, collected over seventeen years, recorded the growth of a cohort of boys and girls correlated with their nutrient intake and used in the construction of an 'Atlas of Growth' which was compared with similar data for British and American children. In other research Cahn undertook analyses of common dietary foods, providing the data used by weight watchers and sportspeople. She was an early proponent of the need to reduce fat intake and analysed the levels of unsaturated fats in meat from different sources. Cahn was a firm believer in a balanced diet as preferable to vitamin supplements, believing in nutrition and dietetics in the prevention of ill health.

Despite opposition from Trikojus, the Diploma in Dietetics was incorporated into the offerings of the Faculty of Applied Science, which was created in 1960 and comprised members from Science, Engineering, Arts, Economics and Commerce, and Law, and one member nominated by the Minister of Education. Its subjects included applied chemistry, electronics, metallurgy and optometry as well as dietetics. The faculty was disbanded in 1969, a year after Cahn herself retired, disillusioned with the treatment of her discipline.

Audrey Cahn was an accomplished painter and sculptor and one of the founders of the Potters Cottage in Warrandyte. She retired first to a cottage which had originally been the gatekeeper's residence of her parents' property in Osborne Road, and subsequently to Canberra where Audrey Cahn Street in Macgregor was named in her memory. She is one of the few Australian biochemists to achieve such a distinction.

A fourth extraordinarily long-lived woman did not survive quite as long as Audrey Cahn, but she shared her artistic interest and talent. Muriel Grace Crabtree (1908–2010) also had a long career as a university administrator. Her association with the department began with her BSc with first-class honours in 1930, preceding its establishment as a separate entity, and stretched to her retirement in 1967: her parallel career with University Women's College (later University College) extended from her time as a resident graduate, tutor and senior tutor from the early 1940s until

Muriel Crabtree, lecturer and university administrator

her appointment as vice-principal from 1955 to 1959. After taking her MSc in 1931, she briefly relinquished her M.A. Bartlett Scholarship in order to take a short-term appointment as a biochemist at the General Hospital in Perth. Her first published paper concerned the relation of gastric function to the chemical composition of the blood.[18] The following year she won a research scholarship to Bryn Mawr College in Philadelphia. From the United States, Crabtree travelled to England with a grant from the Medical Research Council, working in the National Institute for Medical Research at Mill Hill, north of London, and the Lister Institute in Elstree. After a year in the University College Hospital Pathology Department, during which time she published a paper on the sodium content of human erythrocytes with Montague Maizels, she returned to Australia in 1938.[19]

She worked briefly with Trikojus in Sydney before taking up her appointment in Melbourne in 1942. In 1947 she replaced W.A. Rawlinson during his leave and in 1949 Brisbane's *Courier-Mail* reported on a 'short holiday in Brisbane'. 'Speaking for women', the paper informed its readers that:

> Modern girls take their University examinations very seriously. It would be hard to find any of the play-girl type in an Australian University today – certainly not in the Women's College of Melbourne University. Such is the opinion of Miss Muriel Crabtree, MSc, senior lecturer in the Melbourne Women's College … Young and vivacious, Miss Crabtree is not at all the blue-stocking type, yet her specialty, bio-chemistry, is one that makes heavy intellectual demands … Girls have the necessary patience for detailed routine research, in which results may be very slowly visible.[20]

Crabtree was indeed a committed feminist and a dedicated teacher, although she published little after her 1946 paper with Trikojus, mentioned in the previous chapter. This may be explained in part by her work at University Women's College, where she was remembered as an enthusiastic and caring teacher and mentor. She lectured principally to dental students and travelled extensively in Asia, Africa and Europe. In 1952 she visited dental schools in England and Scotland to investigate their teaching of biochemistry to dentists. Muriel Crabtree was also an accomplished painter and long-time member of the Victorian Artists Society. An exhibition of her work was held in 1994 and another at University College in 2011.

As well as Nancy Hosking, there are two more women in the photograph of the Department of Biochemistry in 1948 – Jean Dale and Jean Doig. Jean Dale, later Lottkowitz née Dickinson (1921–1999), joined the department in 1942 as a part-time demonstrator and was promoted to senior demonstrator in 1945. In 1951–52 she spent fifteen months at the Department of Biochemistry, Postgraduate Medical School, Hammersmith Hospital, London. Jean Doig, later Gaze (1924–2012), left the Biochemistry Department in 1952. She had joined in 1945 as a temporary postwar demonstrator in 1945 and been promoted to senior demonstrator in 1950.

Dora Winikoff (1906–1982) did not spend many years in the department, but she was remarkable at this time in its history, first for having taken her MSc degree from a European university (Cracow) and second for continuing to work after the birth of her daughter in 1945. Winikoff joined the department as a research worker in 1941, worked as a part-time demonstrator in 1952, and was a research fellow from 1953 to 1957. Although she undertook some investigations with Kathleen Law in 1943 on vitamin C, comparing blackcurrant juice, tomato juice, fresh tomatoes and dried lemon powder, and published a paper on calcium, magnesium and phosphorus in the milk of Australian women in 1944, Dora Winikoff's principal research interest was in thyroid function and her work in the department was conducted mainly with Trikojus.[21] With him, in 1948, she published 'N^1-Diethylsulphanilamide: a Reagent for the Colorimetric Estimation of Thyroxine'.[22]

While at the Diabetic and Metabolic Research Unit of the Alfred Hospital in the 1960s, Winikoff produced a number of papers. Her last published paper – with Malvina Malinek – on the predictive value of the thyroid 'test profile' in habitual abortion, appeared in 1975 in the *British Journal of Obstetrics and Gynaecology*.[23]

Finally, there was Mary Teresa McQuillan (1918–2004), the woman whom Humphreys tells us Victor Trikojus relied on for her 'sophisticated writing skills'.[24] McQuillan took her BA in 1943 and her MSc (Hons) in

1945, and in 1953 she became the first person to take a PhD from the Department of Biochemistry, with a thesis entitled 'Biochemical Studies on the Thyroid Gland'. In 1948 she spent time at Boston University studying the experimental thyroid work being undertaken there and in 1950 spent a month in New Zealand at the University of Otago. Appointed as a part-time demonstrator in 1944, she became a lecturer in 1953.

In 1958 McQuillan attended the 4th International Congress of Biochemistry and worked at Cambridge

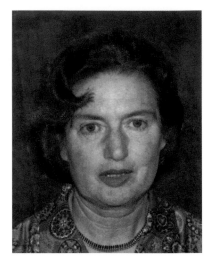

Mary McQuillan, exceptional teacher

University. Reporting on her time there, she noted that Cambridge teaching was of a very high standard but that less biochemistry was taught in the medical course there than at Melbourne. She spent a month at the Isotope School of the Atomic Energy Research Establishment at Harwell, attending lectures in nuclear physics, radiochemistry, radiological protection and the use of isotopes in industry and research, as well as undertaking practical work on the measurement and handling of radioactive isotopes.[25]

In support of his application to have her position reclassified as senior lecturer, Trikojus set out her duties:

1   Lectures and tutorials on endocrinology to Medicine I – assistant examiner;
2   Lectures on endocrinology to final year Science and Agricultural Science students;
3   Supervision of practical work for the Biochemistry section of Physiology and Biochemistry Part I for the Science course and the Agricultural Science course. [In 1960 there were nine demonstrators and 113 students];
4   Occasional postgraduate lectures to medical students preparing for higher degrees.[26]

Mary McQuillan was promoted in 1961 and again in 1974. In the discussions preceding her second promotion, her referees emphasised the excellence of her teaching and the fact that a heavy load, consciously assumed after the departure of Trikojus in 1968, had resulted in a relatively low research output. Recommending her promotion, Joseph Bornstein, who had worked with her in the department, wrote from Monash that:

> The strongest point in Dr McQuillan's favour is her teaching. She is a clear, lucid lecturer, highly conversant with the literature in endocrine biochemistry. Her lectures as I know them are always up to date, and she has the rare gift in a highly complex and often controversial field of not lecturing over the students' heads. At learned society meetings and in research seminars her contributions are always clear, well documented and to the point.[27]

The first and second volumes of McQuillan's *Somatostatin* were published in 1977 and 1979, respectively.[28] She retired from the department in 1982.

Mary McQuillan's time in the Department of Biochemistry spanned its transition from the administration of Victor Trikojus to those of his successors. The women of the department made up a truly remarkable cohort, both for their contribution to various aspects of the discipline of biochemistry and for their sheer longevity.

The Department of Biochemistry in 1948 – *left to right*: *back row* Francis Hird, Jean Doig, Philip Scutt, John Legge; *middle row* Nancy Hosking, Jean Dale, Muriel Crabtree, Audrey Cahn, Jean Millis; *front row* William Rawlinson, Ivan Maxwell, Victor Trikojus, Joseph Lugg, Archibald Gallacher

CHAPTER 4

After the War

A MONG THOSE IN the photograph of the Department of Bio-chemistry in 1948 are two men, Professor Victor Trikojus and Dr Ivan Maxwell, who remained associated with Biochemistry until their retirement. The other, younger men remained in the department for varying amounts of time. For three, it was where they made their professional lives; the other four were to leave for other universities, confirming the reputation of Trikojus's department as a good place to start a career as well as one that provided an environment in which one could undertake outstanding research over a lifetime.

In the immediate postwar period, the department also recruited distinguished scientists to give specific courses of lectures, men such as Paul Fantl (1900–1972) who gave yearly lectures from 1942 to 1952. Fantl's career illustrates yet again the debt Australian science owes to Adolf Hitler.[1] A distinguished biochemist in his native Austria, Paul Fantl left shortly after the Nazi takeover and arrived in Melbourne in 1939 to take up an appointment as organic chemist and biochemist in the Baker Medical Research Institute at the Alfred Hospital, Prahran. He is principally remembered for his research into coagulation of blood.

Australia also benefitted from the emigration of Alfred Gottschalk (1894–1973), who came to Melbourne in 1939. From 1941 to 1951, while working at the WEHI on the biochemistry of yeast, and later with Macfarlane Burnet on the influenza virus, he gave lectures in the department; in 1951 he was also awarded the David Syme Research Prize and elected a fellow of the Royal Institute of Chemistry. He took his DSc from Melbourne in 1959, when he retired, initially to the Australian National University (ANU) and finally to Germany, where he continued his work on glycoprotein chemistry. The Gottschalk Medal is awarded annually by the Australian Academy of Science to researchers in the medical sciences under the age of forty.

Other notable biochemists who delivered lectures during the war and immediately afterwards include Hugh Ennor, the foundation Professor of Biochemistry at the John Curtin School of Medical Research at the ANU from 1948 to 1967, and the permanent head first of the Commonwealth Department of Education and Science from 1967 to 1972 and then of the Department of Science (later Science and Consumer Affairs) from 1973 to 1977. The contribution of Gordon Lennox, who gave several series of lectures between 1942 and 1951, was noted earlier.

As well as employing these stellar outsiders, and Ivan Maxwell as a part-time lecturer, at the end of the war the department was able to appoint additional staff and expand its activities. Some recruits stayed only a short time before promotion to other institutions; others found a permanent home in the department.

One of the most significant appointments of the postwar period was that of William Arthur Rawlinson (1915–1972), who took his BSc in 1938 and his MSc in 1939, shortly after the war began. Rawlinson worked at the WEHI until 1944, when Trikojus appointed him to a senior lectureship in biochemistry. In 1949 he was promoted to associate professor, having previously acted as professor during Trikojus's overseas leave. Rawlinson is remembered in the university both as a biochemist and as an athlete: he was three times the Victorian champion javelin thrower, and the university's Rawlinson Track is named in his memory. He was able to combine his interests when he travelled to England in 1947 on a Wellcome Research Foundation Travelling Fellowship to complete an important piece of research in blood chemistry undertaken in collaboration with the biochemist and poet Claude Rimington (1902–1993) at University College, London. While in England he continued his athletic career, reporting to Trikojus that:

> My athletics are coming to an end. I was pleased to gain 3rd place in the British section of the AAA Champ. And 6th in the all-comers' final. The University of London presented me with full colours for athletics. I was selected to represent England in the International Universities Games in Paris at the end of the month, but must decline with regrets, too much work.[2]

Rawlinson's investigations of the physical biochemistry of those proteins involved in the transport and utilisation of oxygen were combined in his later research with the biochemistry and physiology of exercise. The authors of his obituary in the *University Gazette* noted that:

> The subtlety behind the combination of oxygen with the various haem proteins led him to develop ideas on the perturbation of proteins by small molecules which were considerably in advance of experimental techniques then available. These ideas on allosteric activation and inhibition were subsequently to become very important concepts in biology.[3]

William Rawlinson at the Snobs Creek Hatchery

The obituary also noted that Bill Rawlinson possessed laboratory skills that were no longer common even at the time of his early death: 'He was an elegant glassblower and a fine experimentalist'.

Rawlinson succeeded Peter MacCallum as president of the Sports Union in 1951 and in 1952 he combined attendance at the Biochemical Conference in Paris with attendance at the Olympic Games in Helsinki as part of the Australian delegation helping Melbourne prepare for the 1956 Olympics. In this he was materially assisted by Russell Grimwade, who demonstrated his interest in the Biochemistry staff, as well as the scientific work of the department, when he wrote to Trikojus that the proposed trip:

> has aroused my sympathetic interest because I feel there is a good chance of his experiences to be gained on this trip redounding to the permanent benefit of our institution through the holding of the Olympic Games in this city.
>
> I have pleasure in enclosing my cheque for £250 as a contribution to the fund that makes Prof. Rawlinson's trip possible, and I should be glad if you would pass it on to him as a travelling allowance.[4]

In 1954 Rawlinson was instrumental in the university's invitation to Franz Stampfl (1913–1995) – who had coached Roger Bannister, the first man to run a mile in under four minutes – to take the position of Director of Athletics. Stampfl went on to coach eleven of Australia's Olympians. His previous experience of Australia had been mixed, including as it did transportation from England on the *Dunera*, internment in the camp at Hay and employment in the Australian Army Labor Corps from 1942 to 1946. Stampfl married a Melbourne woman before returning to England at the end of the war, but there appears to have been some hesitation over inviting him to Australia, with Ken Moses commenting in *The Argus*:

> Negotiations to bring Stampfl to Australia have been going on for two years, the man behind the scenes being Professor Bill Rawlinson, of Melbourne University.
>
> The cost of bringing him to Australia will be cut six ways. It will be shared by the National Fitness Association, the Education

Department, the physical education department of the University, the University Sports Union, the Victorian Amateur Athletics Association and the Victorian Women's Amateur Athletic Association.

But already those costs have been underwritten by an anonymous sportsman.

So all the V.A.A.A. has to do is to give its approval to the scheme and Stampfl will be here about mid-July.

If the association decides to haggle over the scheme, everything will be lost.[5]

Rawlinson chaired the Recreation Grounds Committee from 1951 to 1957 and was also active in persuading Sir Frank Beaurepaire (1891–1956) to assist in improving the university's facilities. Sadly, the Beaurepaire Centre named in his honour was not opened until April 1957, well after Sir Frank had died in May 1956. It was, however, completed in time to be used as a training centre for the Melbourne Olympics in November and December 1956.

At the end of 1945 Trikojus wrote to Joseph William Henry Lugg (1907–1997), who is seated on Trikojus's left in the 1948 photograph, in the hope of attracting him to Melbourne as senior lecturer in plant bio-chemistry. Lugg was born in Boulder, Western Australia, and it was to the west that he eventually returned. After graduating from the University of Western Australia, he worked from 1929 to 1945 at CSIR in Adelaide, rising to the position of senior research officer. He was to remain at Melbourne University only until 1948, leaving to become Professor of Biochemistry at King Edward VII College of Medicine and the University of Singapore, after which, in 1956, he took up the position of founda-tion Professor of Biochemistry at the University of Western Australia. There he was succeeded in 1972 by another Melbourne alumnus, Ivan Oliver, who had gone from Melbourne to the University of Sheffield. In Melbourne, Lugg worked on plant proteins with F.J.R. Hird and on the separation of biologically important organic acids. In *Who's Who in Australia* he listed his recreations as reading, travel and archaeology, and his pet aversion as university politics. That said, he marched in protests with students and was even arrested with them, evidently sharing the left-leaning politics of John Legge, who was to join the department in 1948.[6]

In his letter of invitation to Lugg, Trikojus also provided a glancing assessment of F.J.R. Hird, who was to be Professor of Biochemistry for twenty years:

It is intended that the new position should result in a closer liaison with the Department of Botany, but Professor Turner has not yet worked out a syllabus for Plant Physiology and Biochemistry Part I (i.e. 2nd Year course) and in any case it would not be available until 1947. However, he may desire you to give a limited number of lectures (say less than 10) to senior students in his Department … The practical class duties should not be extensive and will be entirely confined to students in Science and Agricultural Sciences. There may be difficult times ahead in the matter of teaching owing to the influx of servicemen, but we are not likely to be greatly troubled next year, and in any case funds can be made available for additional temporary staff. I have endeavoured so far to organise the teaching in my Department so that each member of staff has at least half the week clear for research purposes. I believe I can provide you with a first class assistant in your work in the person of a graduate in Agricultural Biochemistry who has been carrying out research with me for the past few months. He is very keen and reliable.[7]

Although he must have been disappointed at Lugg's short tenure in the department, Trikojus provided a generous tribute to his capacities when he wrote to the Inter-University Council for Higher Education in the Colonies:

In his present position, Dr Lugg has lectured in plant biochemistry and on the chemistry and biochemistry of the vitamins to the senior students in the Faculties of Science and Agricultural Science and to second year students (numbering about 10) in the Faculties of Dentistry, Science and Agricultural Science on elementary general biochemistry. He has also experience in demonstrating to practical classes. His lectures, whether to undergraduates or to graduate societies, are characterized by thoroughness in their preparation; in delivery he is most simulating and must be regarded as one of this

University's best lecturers ... Moreover, Dr Lugg's nimble mind and sound knowledge of the fundamental principles of biochemistry enable him to grasp rapidly the essentials of a problem and to simulate discussions in fields far removed from those in which he has been more intimately associated. It is my considered opinion that he has the qualities which would enable him to occupy a chair of biochemistry with distinction in any part of the British Empire ...

He is a recognized authority on plant proteins ... His laboratory technique is immaculate – I have yet to meet a more careful experimentalist.[8]

Since two of the staff of Biochemistry at Melbourne, Jean Millis and Joseph Lugg, left for positions at the University of Singapore, it is worth comparing the situation of the two institutions at the time. King Edward VII College sent out a frank assessment of its position:

The equipment of the laboratories and clinical departments before the war was probably better than in many of the Medical Schools in Great Britain. Unfortunately serious losses have resulted from the Japanese occupation. The Governments have, however, provided generous grants for re-equipping the College and large orders have been placed ...

There were 265 students in the College in January 1947 ... There were 41 women students ...

The cost of living is high but has been coming down steadily during the past year as supplies have increased. Food is plentiful and is becoming cheaper – the cost at present is slightly higher than in England. The cost of living is unlikely to become stabilized at a reasonable level until the supply of rice is adequate.[9]

The effect of the influx of servicemen and others referred to in Trikojus's letter to Lugg on total University of Melbourne enrolments can be observed from its *Annual Report 1939–1946*. Total university enrolments for 1946 were 6206, compared with 3869 in 1938. In agriculture (including veterinary science) they were 120 compared with seventy-one, and in science there were 956 students compared with a prewar total of 471.[10]

Although the availability of rice was unlikely to trouble members of the department in Melbourne, equipment was still scarce and took many months to arrive from overseas. During her time in America in 1948, Mary McQuillan received a steady stream of requests from Trikojus, asking her to send books back to him and renew journal subscriptions on his behalf. In June 1948, he adjured her, 'Don't fail to get details of the Barker method for the estimation of iodine. If the equipment necessary is unusual, you had better try to bring back a set with you if the funds will allow it. I have already suggested that you should bring some ceric ammonium sulphate'. A month later, he asked, 'When in Pasadena please buy a rubberized cover for the Beckman spectrophotometer (National Technical Laboratories, South Pasadena)'.[11]

Scarcity of equipment also featured in early correspondence between Trikojus and another short-term staff member who was to attain a degree of notoriety after he left the department. Peter Henry Springell (1927–) applied to the professor after graduating BA with first-class honours in biochemistry from St Catharine's College, Cambridge and working with Michel Macheboeuf (1900–1953) at the Institut Pasteur in Paris on lipo-proteins. His appointment as a demonstrator for three years from 1949 was slightly complicated by a bizarre letter from Professor Albert Charles Chibnall (1894–1988) alleging that Springell was in fact a fraud whose real name was Sprinzels. This was refuted by numerous testimonials, including one from Macheboeuf himself, allowing Trikojus to appoint Springell and ask for a favour. On 4 November 1948, he wrote:

I am glad to learn that you have been able to book a passage from January of next year. This should bring you to Melbourne in ample time for the beginning of the academic year ... Let me know if there is anything of a special nature which you would wish to have ordered through the University's agents in London, Messrs Jepson, Bolton & Co. Ltd ... or which you would wish to bring with you. There is one item on which the University's agents cannot get any satisfac-tion from the French manufacturers (Messrs Durieux, 18 rue Pavé, Paris), namely the supply of Chardin filter papers of the following quality and dimensions: 50 cm Chardin Fluted Filter Papers (already folded). If you could manage to obtain some of these (say 1 gross) and bring them with you, it would be a great help to work proceeding in the Department.[12]

Trikojus was keen, however, to make clear that the department could cater for the needs of the new recruit, who had requested a Kumagawa lipid extractor. In December 1948, he asked Springell to send a sketch of what he wanted, telling him:

> We have excellent facilities at present for glass-blowing, as the University has recently appointed Mr Kamphausen who was formerly in charge of the glass instrumentology section of the I.G. Farbenindustrie at Elberfeld, Germany. Also, we have good workshop facilities in the University. We have in the Department an all-metal extractor which can handle 1 kg of solid material and have, in addition, all-glass continuous extractors for use with solvents heavier than and lighter than water.[13]

Springell, working with Francis Hird, took his PhD from Melbourne in November 1953, by which time he was already working with CSIRO

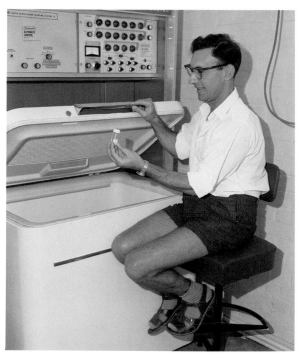

Peter Springell, senior research scientist at the National Cattle Breeding Station, Belmont, at a liquid scintillation spectrometer at CSIRO's breeding station

because Trikojus had been unable to fund a position for him. Writing in June 1953, the professor told the registrar of Adelaide University that Springell was 'an indefatigable worker, spending long hours in the laboratory and his technique is good'. Peter Springell rose to become a principal research scientist with CSIRO in Rockhampton, where he undertook comparative research into the genetic attributes important for productivity of cattle in the tropics. He left in 1976 following a controversy about his public criticism of the organisation's research priorities.[14] He later worked in the Northern Territory.

The small staff of the immediate postwar period had also to turn their hands to domestic tasks. On 11 July 1950, Trikojus wrote to the registrar of a working bee held the previous Saturday among teaching, research and technical staff to clean the extensively blackened area of the ceiling and walls of the research section of the department resulting from the fire mentioned earlier. He submitted a list of incidental expenses amounting to a total of £3.1.6 for reimbursement of £1.10.7 spent on lunch and teas, £1.3.0 on gloves and 7/11 on a broom. He also noted that Archibald Gallacher and Peter Hunter – who had evidently recovered from an accident in 1948, in which he fell from a tram on his way to work and suffered a fractured skull – 'did exceptionally good work, partly in their own time, in clearing up the damage caused to equipment. In this way quite a considerable saving was effected'. He recommended payment of 3 guineas to Gallacher and £1.10 to Hunter.[15]

The 'first-class assistant' recommended to Lugg by Trikojus was associated with the department for over four decades. Francis John Raymond Hird (1920–2014) took his BAgrSc in 1945, winning the J. M. Higgins Exhibition and his MSc in 1947. Hird had not come to the university directly from school, which he had left at the age of fourteen. He had worked as a messenger in the Department of Customs, and, after studying for his Leaving Certificate at Taylor's College, as a primary school teacher. He enrolled in agriculture in 1941, passing all four first-year subjects at the first attempt. The *Annual Report 1939–1946* recorded his research with Lugg:

> Leaf proteins – whole leaf protein samples have been prepared from the foliage of *Erodium cygnorum* growing on soils believed to be

(a) healthy and (b) copper-deficient. Partial amino-acid analyses of the preparations are under way. For use in analyses of plant proteins, special apparatus has been constructed and reagents have been prepared.[16]

In 1948 Hird published 'Paper Partition Chromatography with Thyroxine and Analogues', and left later in the year to pursue postgraduate work on peptides at the Sir William Dunn Institute of Biochemistry at Cambridge University with E.V. Rowsell.[17] He took his PhD on transamination and transpeptidation reactions from Cambridge and returned to Melbourne in 1951 to take up a senior lectureship in agricultural biochemistry. In 1953, on a Fulbright-Smith-Mundt Award, he studied with Joseph S. Fruton (1912–2007) at Yale. Promoted to reader in agricultural biochemistry in 1957, Hird spent 1960 at the University of Oxford and at Carlsberg Laboratories in Denmark. He was awarded his DSc for a collection of published papers in 1962. The second Chair in Biochemistry, to which Hird was appointed, was established in 1964. The *University Gazette* noted at the time: 'In creating a second Chair, the aim has been to extend the development of Biochemistry in accord with the

Francis Hird in Cambridge

increasing sub-divisions of this scientific discipline, rather than to create a separate department within the School of Biochemistry'.[18]

Although the discipline of biochemistry at the University of Melbourne owes its foundation to the establishment of a Faculty of Agriculture, the occasionally uneasy relationship between Biochemistry and Agriculture was exemplified early in Hird's time in the department by tensions with Trikojus over the acceptance of grants from the wheat and wool industries, for which Hird applied from the Faculty of Agriculture. He was awarded £10 000 a year from the former, as well as a 'fairly substantial' grant from the Rural Credits Development Fund of the Commonwealth Bank, which enabled him to purchase his own equipment without reference to the Head of the Department of Biochemistry. Trikojus, with his preference for central control, found this difficult to accept.[19]

In addition to his own research, ranging from studies on proteins and enzymes, to studies on cell organelles, whole cell preparations, organs and whole animals, the Academic Board noted on his retirement:

> Frank Hird has had an outstanding career as a teacher, both at undergraduate and postgraduate levels. Many postgraduate students have had cause to reflect on his ruthless pruning of their manuscripts so that the essence should emerge … All of them must have profited by his insistence on clear, controlled technique, on straight-forward prose … He insisted that everything they did or wrote must be based on reason and knowledge. He lectured always without notes of any sort … As he grew more experienced in lecturing, he was able to indulge in a few eccentricities – his great weakness, the limerick, could sometimes be heard in the lecture halls.[20]

The following, recollected by a medical student, can serve as an example:

> There was a young man from Siluria
> Who slept with a girl from Manchuria.
> It wasn't the clap
> That got that young chap
> But haematoporphyrinuria.

Hird was Head of the Department of Biochemistry from 1968 to 1974, Dean of Science in 1972 and an active member of many university committees. He was also a regular and outspoken commentator on general university policy, with letters to the *University Gazette* occasionally sparking an equally outspoken response. In 1999, as noted earlier, Francis Hird and Max Marginson conclusively rebutted the assertion that Trikojus had demonstrated Nazi sympathies.[21]

Although it was written before his elevation to a chair and before he became Head of Department, Hird's report to the University Council on his return from study leave in 1961 provides a vivid foretaste of some of his concerns with the administration and funding of his discipline. Noting that his time had been split evenly between two institutions, and that he believed that it would have been more profitable to have devoted the whole ten months to work in one laboratory, Hird dismissed the research itself in two sentences: 'This time was spent investigating certain aspects of the reactivity chemistry of the disulphide bonds of serum albumin and insulin. The investigation was completed and the results will be submitted for publication'.[22]

The remainder of his report concerned the increasing availability of electronic instruments for analytical procedures. The increase in funding that would be required to acquire these and various other factors he believed necessitated a reassessment of methods of training of all students of biochemistry.

As further biochemical knowledge unfolds, the more physical and mathematical become the concepts. In addition, the more rapid the rate of development of biochemistry, the sooner an individual's fundamental knowledge becomes inadequate. It is time that this question was given serious thought, at both the post-graduate and undergraduate level. In this connection, the responsibility is partly my own as I cannot remember any discussion between myself and the teachers of our first year basic subjects of biology, chemistry and physics. The time is here for a reappraisal to be made.[23]

He concluded by making a scathing comparison with a risk-averse mentality that starves good research of support for fear that 'some people of low merit may get too much money' and commended the American system under which 'No good person lacks all the money he needs to let him follow his nose in his research problems'.

Joining the department in 1948, at almost the same time as Hird, was another graduate of the University of Melbourne, one who had spent much of the war in physiological research on mustard gas with the Australian Chemical Warfare Research and Experimental Section of the Services and Munitions Department in Queensland. John Williamson Legge (1917–1996) had taken his BSc in absentia in 1938 because he had already left Melbourne to work with Max Rudolph Lemberg (1896–1976) at the Kolling Institute of Medical Research in Sydney. Their very successful *Hematin Compounds and Bile Pigments* was published in 1949.[24] He took his MSc from Melbourne in 1946 with a collection of previously published papers, including work on liver catalase, haemoglobin and ascorbic acid, and a new preparation of biliverdin.

Jack Legge, who worked on mustard gas during World War II

The work on mustard gas was undertaken with Hugh Ennor and together they designed an air-conditioned gas chamber measuring 100 cubic metres. Jack Legge's Melbourne colleague, Max Marginson, made clear the terrible choices facing scientists in wartime:

> His job was particularly unpleasant, since it involved exposing volunteer soldiers to mustard gas in tropical conditions. But as Jack said, there was information available that the Japanese had large stocks of this gas and that it was more potent in tropical conditions than it had been when used in France. He told me that, grisly as this titration of tolerance proved to be, in view of the times, the necessity for the experiments was absolute.[25]

A 1948 letter from Trikojus to Legge suggests that, although he would no longer have to design his own equipment, it might not always be easy to come by:

Regarding the cold chambers which were installed by the C.S.I.R. for Professor Young's use, these are unfortunately in a laboratory which is now being used by Physiology for overflow students in that Department. Physiology objects strongly to our using the chambers owing to the noise made by the motor and relays. However, if they prove to be essential, I shall arrange to have the motor, tank and at least two of the cabinets moved to a section of Biochemistry where the noise will raise no objection.[26]

The letter continues:

There will be no difficulty about your giving a few lectures, in fact Burnet is in favour of it and so are other people on the Medical Research Council. I have indicated in my previous letter a couple of things which I should like you to help us with this year. I am enclosing a copy of our lecture course in the third year which I hope is self-explanatory.[27]

Jack Legge's interest in mustard gas — a research area also pursued by E.R. Trethewie of the Department of Physiology and several other members of various departments of the university during the war – extended to lobbying for servicemen and volunteers affected by the trials who developed health problems later in life as a result.

Legge was active in the Communist Party from 1935 and on 29 October 1954 he was called before the Royal Commission on Espionage (Petrov Commission). He adamantly denied spying or providing information to 'anyone not in the service of the Crown'.[28] His cousin, George Williamson Legge, who had taken his BA (Hons) from Melbourne University in 1937 and won the Laurie Prize for Philosophy, and who had been a staff member of the Department of External Affairs from 1944 to 1953, was questioned at the same inquiry. A meeting of the two cousins in Canberra came under particular scrutiny. During this time it was notable, although he could never have been thought to share Jack Legge's views, that Victor Trikojus went out of his way to shake his colleague by the hand in public.

In the department, Legge established a cellular biochemical unit and reported in his first year that large quantities of plasma had been treated for the production of plasma cholinesterase with the object of purifying this enzyme and studying its properties in detail. With CSIRO, his team also investigated the mechanisation whereby certain pseudomonads isolated from arsenical cattle dips are able to oxidise arsenite to arsenate. During a study tour in 1958–59, Legge attended the International Biochemical Congress in Vienna, spent a term at the Laboratoire de Morphologie Animale of the Université Libre in Brussels and visited biochemical centres in Europe, including the United Kingdom. Following the tour, Legge reported to the University Council that this was an interesting time to be in Europe, with a referendum and elections in France, the announcement of education reforms in the USSR, the advent of the European common market and the last weeks of the International Exhibition in Brussels. He devoted much of his report to educational reforms in various countries, writing:

> The basic reason behind these educational reforms is the recognition that the various types of school practical work and the sixty to eighty hours of practical production work per year, which the eighth to tenth forms do, may be adequate to acquaint the pupils with various categories of adult work but are 'far from marriage of education to productive labour' which is desired …
>
> We are, I think, in Melbourne, able to give the medical students a more extensive course in their pre-clinical years than I have found in many universities overseas. On the other hand, our shortage of staff and lack of proper facilities for tutorials and discussions with students, while perhaps not as serious as in France, are still formidable barriers to the improvement of our course.[29]

His own teaching was very highly rated, with Marginson commenting on his development of 'a lecture course to Medical Students of astonishing percipience', including a 200-page lecture synopsis which was 'up to the minute in its text and was full of remarkable biological insights. Perhaps it was over the students' heads, but it was certainly the best text in Australia and was eagerly sought after by others'.[30] Legge's publications ranged from the presidential address to the Melbourne University Science Club in 1950 to papers with A.W. Turner on the bacterial oxidation of arsenite and another on the relevance to colorectal screening of peroxidase levels in food.[31] In a department in which the distinction between academic and non-academic staff was observed to the point of having separate sittings for morning tea, Legge was notable for his reluctance to pull rank. Timothy Anning, on one of his first days at work as a very junior technical assistant, was told not to let people into a particular laboratory. He refused entry to Legge, who philosophically and without discussion simply went downstairs, along a corridor and let himself in through another entrance. Jack Legge is also remembered for infrequently wearing a lab coat and having a predilection for snuff, with which his fingers were stained and his clothes often sprinkled.

Another member of staff who spent almost as short a time in the department, and left for a stellar but sadly truncated career elsewhere, as Joseph Lugg had done, was Robert Kerford Morton (1920–1963). He came to the department as a senior lecturer in 1952 and was promoted to Associate Professor of Plant Bio-chemistry at the beginning of 1957, leaving the same year to take the Chair of Agricultural Chemistry at the Waite Research Institute of the University of Adelaide. Trikojus, who wrote that he had managed to induce the University of Melbourne to create a position specifically for Morton (whom he had met at Cambridge the year before), paid a handsome and detailed tribute to a highly esteemed colleague and friend, which gives more information than can be provided here.[32]

Robert Morton, a brilliant post-war recruit

Morton, dux and gold medallist of the Hawkesbury Agricultural College, from which he took his Diploma in Dairy Technology in 1938, enrolled at Sydney University in 1939, interrupting his studies in 1941 to join the Royal Australian Navy as a submariner: by the time he was demobilised in 1946, he had risen to the rank of lieutenant-commander. This experience seems to have left its mark in more ways than one. Trikojus noted that Morton had endured danger and personal discomfort: Humphreys records that Dr Morton expected people to stand when he entered the laboratory.[33] Graham Parslow notes that he expelled an Adelaide student from his lectures for not wearing a tie.[34]

Morton took his BAgrSc in 1948 with first-class honours and the University of Sydney Gold Medal. Having been awarded the inaugural Gowrie Travelling Scholarship, he went to the School of Biochemistry at Cambridge, taking his PhD in 1952. A related paper, in which he described the application of the purification method he had devised to eleven distinct enzymes, including a number of phosphates and succinic dehydrogenase and phosphates, appeared in *Nature* in 1950.[35]

The obituary tribute by Trikojus summarises the poignant end to 'five years of almost feverish activity' at Melbourne and Morton's later work:

Perhaps the contributions for which he will be best remembered, in the annals of biochemistry, lie in a series of brilliant studies of the enzyme, yeast lactic dehydrogenase, which occupied a considerable part of his time in the years 1953–1963. It is sad to recall that it was while working on an improved method for the purification of this enzyme, involving large quantities of acetone, that an explosion occurred which resulted in his fatal injuries.[36]

Morton was elected a fellow of the Australian Academy of Science in 1957 and was one of the organisers of the International Union of Biochemistry symposium on haematin enzymes, held in Canberra in 1959.[37] He was active in the establishment of the Australian Biochemical Society and was its president from 1958 to 1959. In 1961 he was president of Section N of the Australian and New Zealand Association for the Advancement of Science. His presidential address, which Trikojus described as 'brilliantly conceived and delivered', was entitled 'New

Concepts of the Biochemistry of the Cell Nucleus'.[38] In his obituary, Trikojus provided a complete list of Morton's publications.

Joseph Bornstein (1918–1994) spent only two years as a senior research fellow in the department. He had taken all of his academic qualifications from the University of Melbourne, being awarded his MBBS in 1941, his MD in 1944 and his DSc in 1956. Bornstein enlisted in the Australian Army Medical Corps after his residency and from 1942 to 1946 worked mainly in Cairns with the Land Headquarters Medical Research Unit on trials of synthetic antimalarial drugs. He turned his attention after the war to diabetes, when he worked with Basil Corkill (1898–1958) at the Baker Research Unit. In 1949 he and Phyllis Trewalla became the first to measure insulin in the plasma of diabetic and normal humans. After working at King's College Hospital, London with R.D. (Robin) Lawrence

Joseph Bornstein studying results obtained during work on a diabetes drug

and Charles Gray, and at the Washington University Medical School in St. Louis with Carl and Gerty Cori, Bornstein spent the years 1953 to 1956 at the Department of Biochemistry as a research fellow. He was the first director of research at the newly established Diabetic and Metabolic Unit at the Alfred Hospital and the foundation Professor of Biochemistry at Monash University from 1961 to 1983.

Francis Hird was succeeded as Head of the Biochemistry School in 1974 by C.A.M. Mauritzen (1927–2003), AHWC, ARIC, PhD (Edinburgh), who had arrived in 1959 from England to take up an appointment as senior research officer. Mauritzen was promoted to senior lecturer in 1962 and reader in 1967, and he served as head of the school until 1980. The *Research Report* of 1966 recorded that he was pursuing structural studies on histones in collaboration with H. Busch, W.C. Starbuck and C.W. Taylor while on leave at the Department of Pharmacology of Baylor University in Houston, Texas: 'Using methods developed at Melbourne, highly purified samples of two of the arginine-rich histones (ß6 and ß7) were prepared from normal and malignant tissues ... A long term objective of the work is to determine the extent of cell specificity of histones and their connection with cell differentiation.'[39] Harris Busch paid tribute to this work in his presidential address to the 1990 Annual Meeting of the American Association for Cancer Research.[40] Mauritzen retired in 1989.

The contribution of another male academic had an effect on the university well beyond the Department of Biochemistry. Maxwell Arthur Marginson (1928–2002) was a polymath with a distinct bent towards collegiality and conviviality, which contributed to his devotion to teaching rather than research. A lively account of the early years of what would be his association of over forty-five years with the department and over fifty with the university can be found in *More Memories of Melbourne University*.[41]

Max Marginson, younger brother of Ray Marginson, the Vice-Principal of the University of Melbourne from 1966 to 1988, came to the university through the state school system and, having won the First Year Exhibition in Zoology, studied biochemistry. Having completed his BSc in 1947, he enrolled for the MSc under Jack Legge, in whom he found a sympathetic colleague on both a scientific and a political level. Although he had a profound influence on teaching within the department,

Marginson's brother noted that: 'To the despair of his friends, he completed his PhD research but was indifferent to the final write-up or the qualification itself'.[42]

He worked as a demonstrator and senior demonstrator in the department from 1948 to 1957, a lecturer from 1958 to 1971, and a senior lecturer from 1972 until his retirement at the end of 1993. He was a foundation member of the Australian Biochemical Society and a Victorian state representative in 1957 and 1958. His initial research was directed, with Jack Legge, to the isolation of secretin and it produced one internationally competetive paper on metabolism. His doctoral research, under Francis Hird, focused on whole tissue and animal biochemistry. This work resulted in two papers co-authored with Hird: 'Formation of Ammonia from Glutamate by Mitochondria' and 'The Formation of Ammonia from Glutamine and Glutamate by Mitochondria from Rat Liver and Kidney'.[43] Both these research projects were adequate to meet the requirements of MSc or PhD but he did not submit, even though very little extra effort was required. Max was rather disillusioned by this aspect of science from the secretin work because Jack Legge shared their very successful unpublished data with their Swedish colleagues, only to find this helped the Swedes to publish first. There was also a great deal of social friction over the authorship of this paper, because Marginson was not acknowledged as the lead author. Within the department, however, Max Marginson is primarily remembered as a teacher and organiser of teaching. He contributed to debates on teaching as early as 1948, setting out the need for national coordination in the teaching of science in Australian universities.[44] His abiding interest in metabolism and nutrition and his determination to disseminate his conclusions to a wider public are

Max Marginson in University House

evident in a 1983 contribution to the *University Gazette*, in which he noted the lack of general knowledge of human nutrition among members of the 'supermarket society' and asserted that:

> It is ironic that research into animal nutrition has been undertaken for decades, but it was only in 1974 that the CSIRO's Division of Human Nutrition was established, and in all the universities in Australia there are only two Professors of Nutrition.[45]

Max Marginson, as both his own writings and his obituary by his brother make clear, had a vigorous intellectual and social life beyond the department. Like many others, he was intimately involved in University House, the staff club of which Victor Trikojus had been president in 1958; Marginson was a foundation member, convenor of the wine-tasting panel, vice-president three times, and president in 1970, 1971, 1983, 1984 and from 1990 to 1992. He was a keen supporter of the Matthaei Collection of Early Glass displayed in the club. As well as writing for publications as varied as the Melbourne University Science Club's *Science Review*, *Australian Naturalist* and *Australian Book Review*, Marginson organised a monthly luncheon meeting of wine producers and consumers and was responsible for the establishment of the Jimmy Watson Trophy awarded at the Melbourne Royal Show. He was also a talented clarinettist and jazz enthusiast, having established a jazz society at Melbourne High School.

Women were, of course, not absent from the department during the period covered here. One woman who spent almost forty years in Biochemistry was particularly influential. Pamela Ellen Emina Todd née Purcell (1925–1997) was first employed in 1947 as a demonstrator in chemistry in the university's short-lived Mildura branch. The branch was officially opened on 17 May 1947 by Victorian Premier John Cain, and its historian J.S. Rogers notes that those attending the ceremony included the Chancellor, Vice-Chancellor and other representatives of the University of Melbourne; representatives of the state government and the city and shire of Mildura; local residents; and members of the branch staff.[46] A special conferring, the first such ceremony in Victoria to take place outside Melbourne, was held and Pamela Purcell received her BSc.

Eleven other junior members of staff graduated at the same ceremony, which was followed by a reading of George Bernard Shaw's *Arms and the Man*.

Pamela Todd rejoined the university staff in 1951, working as a research assistant for two years before leaving for England, where she spent 1953 to early 1956 as a research officer at the National Institute for Research in Dairying at Shinfield, near Reading. She was a Nuffield research scholar from 1956 to 1959 and then a lecturer in biochemistry until her retirement. In 1970 she was seconded to be assistant to the Dean of Science and in 1978 she took overseas leave to work with W.A. Gibbons at the College of Agricultural and Life Sciences at the University of Wisconsin, Madison. Her report to the University Council of 31 October 1978 made clear the tension so often experienced, notably by women, between research interests and teaching commitments:

Pamela Todd, lecturer and college tutor

> Our main interest now was to find out if the enzyme which hydrolysed the synthetic peptide was also capable of hydrolysing the equivalent bond in a peptide or more specifically in a pro-hormone to give the hormone. To investigate this we first needed a pure enzyme since the crude pituitary extracts we were using were of course a mixture of many different proteins including a whole range of peptidases.
>
> We were able to substantially purify the enzyme relatively easily due mainly to the fact that we were not restricted in our strategies by the lack of funds. To an Australian the amount of money available for day-to-day running expenses was really unbelievable ... leave expired and I was compelled to return to Melbourne and to my third term teaching commitments.[47]

Pamela Todd was vice-chair of St Hilda's College from 1980 to 1984, having been the foundation senior tutor for the previous ten years. She was a foundation member of University House and a member of the

Lyceum Club from 1948 to 1960. She retired from the university at the end of 1984. In 1989 St Hilda's College established the Pam Todd Research Scholarship, which is awarded in her honour to a resident member of the Senior Common Room who demonstrates outstanding merit and promise in his or her scholarly discipline.

The department could not, of course, function without technical staff, and in this period, one man, apart from a three-year stint in the RAAF, was to work in Biochemistry for twenty years, from 1941 to 1961. Before this, when the Department of Biochemistry was established in 1938, John Egan (1889–1951), known to the students as 'Hooks', had been placed in charge of the biochemistry laboratory. He had been appointed to Physiology in 1911 and promoted to technical assistant in 1923. Russell tells us, 'An inveterate smoker, he was always to be seen with a cigarette stuck to his lower lip'.[48] John Egan transferred to the Maintenance Department in 1943.

This was the same year that John Tucker Ford (1924–1997) left after two years with the department to join the RAAF. He had been appointed to Biochemistry in 1941 as a laboratory boy and promoted to laboratory assistant the same year. When he applied to enlist in the air force, Ford stated that he had served as a staff cadet at the university, and his RAAF personnel record notes that he spoke German and, somewhat unexpectedly, Latin, as well as having a good knowledge of first aid. He completed the Medical Orderlies Course in 1943, gaining a mark of 85 per cent for subjects including anatomy and physiology, medical and surgical nursing, ward management, patient hygiene, first aid, theatre instruments and RAAF organisation. On his discharge in June 1946 he was stationed at No. 4 RAAF Hospital in Sale. John Ford rejoined the department in 1946 and was promoted to technical assistant B in 1947.

It is not untypical of the records of technical staff that their activities are recorded in official documents principally when they cost the university money, through notification of their salary increments. Minor accidents are also recorded, hence the Trikojus papers note that on 8 August 1946 Ford was injured when he collided with another technician who was carrying a glass pipette, which broke and cut Ford's hand, and that on 22 March 1948 he was slightly injured in an explosion while assisting a plumber to trace a gas leak in the main students' laboratory.

November 1950 saw a less serious, though common, mishap when Ford is recorded as having ruined his trousers with cleaning fluid, for which he was reimbursed £2. When John Ford resigned in 1961, it was as a senior technical officer grade 2.

One other entry in the Trikojus papers from the period immediately following the war serves to illustrate the effects of shortages on the work of the department. A report to the registrar on 18 July 1947 stated:

> A.W. McEachern, Workshop technician, has placed his car at our disposal for essential services which cannot be provided by one of the juniors with a bicycle … I recommend an allowance to Mr McEachern of 4d per mile, but at the same time he should not be required to use his personal petrol ration tickets for this purpose. Would it be possible to make an issue of tickets to cover 4 gallons per month, this to be regarded as a maximum, and on the understanding that unused tickets would be returned to the office?[49]

The generous man who lent his car was Archibald William McEachern (1917–1998), who joined the department in 1945 having served the previous four years in the No. 1 Flying Personnel Research Unit of the RAAF. He resigned in 1949 despite the efforts of the professor to improve his working conditions. In July 1949 Trikojus had told the registrar:

> Mr McEachern, who is in charge of the Workshop administered by my Department, but serving also the interests of Bacteriology and Pathology, has given notice of his intention to vacate his position at the end of August. According to McEachern, his sole reason for offering his resignation is financial. As a gesture to McEachern, I desire to recommend that he be offered the higher grading of Senior Technician, with an immediate increment at the rate of £50 per annum. In making this recommendation I understand I have the full support of Professor MacCallum whose Department, like mine, has gained considerably from the workshop facilities which McEachern has built up during the past four years. We should keep in mind also that it will be almost impossible to replace Mr McEachern adequately at the salary which he is at present receiving.[50]

In August McEachern wrote to Trikojus from Albury, where he was doing electrical work, that business was very brisk, as his firm was servicing the greater part of the Kiewa Valley up as far as the Bogong Hydro Electric Scheme.[51] The electrical contracting firm that Archibald McEachern established in Wodonga in 1950 is still operating. Now known as D.N. Bishop and Co. Pty Ltd, it was headed from 1983 to 2002 by his son Malcolm, after which time his grandson Gregory took over.

Archibald McEachern, workshop technician

There was also an army of men and women who kept the department clean and tidy. Throughout its history, the university has depended heavily on the services of men like William Henderson, who washed its linen, and women like Bridget Peck, who cleaned the floors and tables of the Department of Physiology during its early years.[52] Some of the conditions of employment in the department can be inferred from the professor's correspondence. In December 1950, he wrote to the accountant:

> None of the cleaners is willing to work in the Department during the first three weeks of January subsequent to the Christmas recess. I have explained to them that it is essential to have at least one cleaner in the Department during this period, and if agreement could not be reached in this regard it would be necessary to obtain a new cleaning staff … In the case of the senior cleaner … she has been invariably accommodating in previous years, and has, this year, what seems a reasonable excuse.[53]

The 'reasonable excuse' was that she had injured her hip in a fall on the stairs while carrying out her duties in March 1948 and spent time on sick leave.[54]

Joyce Calvert (1924–) spent thirty years in the Department of Bio-chemistry, working as a cleaner and bottle-washer from 1956 to 1986. During that time the cost of clean-ing services to the university was frequently a matter of earnest consid-eration by the university management, in terms that might be described as less than generous. The arguments for and against contract cleaners have a long history, but it is certain that, like the academic staff, many of the university staff, including, for a variety of reasons, many of its 'General Staff', chose to remain for many years.[55] Those who

Joyce Calvert, cleaner and bottle-washer for thirty years

did not may have left for higher pay, but their work was not unappreci-ated – the chief technical officer, Lindsay Rayner, noted the high standard of cleaning in the class laboratories, especially when the requirements were clearly enunciated and cleaners were treated as part of the team. When one cleaner left in 1952, W.A. Rawlinson, acting Head of Depart-ment at the time, wrote to her: 'As you have given such good service to us, I thought that you might like a reference even though you have not requested it'. The reference read in part: 'She gave most conscientious and loyal support. It was a pleasant surprise to find in her a worker who was willing to do something more than a superficial job. I give my strongest recommendation to any future employer who requires a trustworthy and efficient Cleaner'.[56]

A great deal of Trikojus's time in the immediate postwar period was taken up with administrative concerns, prominent among them the effort to see the department established in new and adequate premises. This goal, eagerly awaited, was not achieved for over a decade.

The Department of Biochemistry, 1978

CHAPTER 5

The Great Shift

Two shifts, each of a very different order of importance, had a considerable impact on the study of biochemistry at Melbourne. The first was the discovery of the structure of DNA, which forever altered the science of biochemistry. The second was a more mundane shift – that of the department to its long-awaited new quarters in the Russell Grimwade Building. The work of James Watson, Francis Crick, Maurice Wilkins and Rosalind Franklin made possible the science of molecular biochemistry: it was not, however, until 1995 that the new name of the Department of Biochemistry and Molecular Biology appeared in the *University Calendar*. Moving into a new building, assembling the whole Department of Biochemistry under one roof, changed its academic dynamic.

Unlike Russell Grimwade, the department's great patron, Victor Trikojus not only lived to move his department into the new building that had occupied so much of his time and effort over the preceding decade, he also worked there until his retirement at the end of 1967, when he resumed bench work as Honorary Professor in the Department of Chemistry. With Grimwade himself dead, it was his widow, Mab Grimwade, who was present at the opening.

A biography of Russell Grimwade was published almost ten years after his death as the first volume in the Miegunyah Press Series, the establishment of which, under the auspices of Melbourne University Press, was made possible by his bequest.[1] It records that Wilfred Russell Grimwade (1879–1955) was one of nine children of a successful whole-sale druggist. He graduated BSc in 1901, marrying Mabel Louise Kelly (1887–1973) in 1909. He bought their Toorak house, 'Miegunyah', the following year as a wedding present for his wife.

Grimwade and his brothers continued to develop the thriving industrial business they had inherited, with Russell taking a special interest in the extraction of oils from Australian indigenous plants (including Bosisto's famous eucalyptus oil) and the establishment of the Australian Oxygen Company, which mutated in 1935 into Commonwealth Industrial Gases. Grimwade's interest in forestry and forest products was wide and deep, ranging from conservation to exploitation. He supported the establishment of the School of Forestry in Canberra and wrote extensively on eucalypts. He served on numerous boards and committees, including that of the WEHI, the Melbourne Botanic Gardens, the National Museum of Victoria and the University Council. Russell Grimwade was a council member from 1935, Deputy Chancellor from 1941 to 1943 and chaired the Buildings Committee until 1953. He declined to put his name forward as Chancellor when Sir Arthur Lowe retired, partly because of his own health and partly because he had himself supported the appointment of Sir Arthur Dean.

Grimwade donated £1000 in 1938 to the Engineering School and £1000 in 1940 to the Chemistry School. He gave £2000 to the Wilson Hall Fund and £2000 towards the furnishing of University House as well as £5000 to the Centenary Appeal. Altogether, his benefactions totalled about £65 000. He also, as detailed earlier, gave smaller sums towards the expenses of individuals, such as W.A Rawlinson's trip to Helsinki.

Both Grimwade and his wife took a keen interest in all aspects of the university's projects. The 1944 gift of £50 000 – paid in annual instalments until 1947 – towards the building to house the School of Biochemistry was the largest made in his lifetime. It was not, of course, his only gift to the department. Although Grimwade's support, once pledged, was unequivocal, he had initially thought of endowing a new and desperately needed library for the university. The generous support of the Baillieu family for that project, however, inspired the Vice-Chancellor, Grimwade's old friend John Medley, to suggest a building for Biochemistry instead. Debate over the location of the new building went on for years, with Grimwade writing somewhat plaintively to Medley on 13 December 1948:

My Dear Jack,

BIOCHEMISTRY

Don't you think the time has come when we can contemplate erection of this School?

The years are passing and I would like to feel that the scheme had been started, and to play some part in its planning and construction.

Poor Trikojus has his department ranging over four separate buildings at the present time, so that it is almost impossible for him to exercise proper supervision of the work done in his department or any control under any but extravagant circumstances.

Sketch plans of this building were made by Everett some couple of years ago. These had Trikojus's and my approval, but, at the time, it was impracticable to proceed with them, and now the nebulous master-plan of Lewis's is a new bogey that confronts us in the completion of this building. Everett's sketch plan showed the building with its back to Tin Alley, and I see no reason why anything but a slight modification to that plan could not make it conform with Lewis's line of buildings which he shows on the extreme northern boundary of our land. Do let us do something about it soon ...

What about a nice little resolution for the New Year: "let us have a new School of Biochemistry".[2]

Two years later, the Dean of Science optimistically, and as it turned out, erroneously, told the council:

The site for the new School of Biochemistry has been re-determined as that between Old Chemistry and Zoology. Active steps are now being taken towards the detailed planning of the new School and it is hoped to make a start with the construction within twelve months. The Vice-Chancellor has obtained promises from the Government towards supplementing the funds available from Sir Russell Grimwade and Messrs Nicholas Pty Ltd.[3]

Yet another year later, Grimwade told the council that if a Medical Centre was to be developed in the south-western corner of the grounds, the Biochemistry Building should be part of this group. If no such centre was to be developed in the next twenty years then it would be better to build it on Tin Alley. The Chancellor hoped that work would finally begin on a site in the south-western corner of the grounds in April 1956.[4]

Finding an architect proved as problematic as finding a location for the building. The initial choice was the chief architect of the Public

Russell Grimwade, businessman, scientist, philanthropist and keen photographer

Works Department, Percy Everett (1888–1967), who had been responsible for the Frank Tate Building, Old Commerce and the Engineering Workshops.[5] His involvement became controversial, however, when objections were raised to a public servant devoting time to a project that was being privately funded. The university decided to employ the firm of Bates, Smart & McCutcheon, who had designed the Buckley & Nunn building in Bourke Street and were to provide the master plan of Monash University.[6]

There could be little doubt of the need for new quarters. Forty years after the first stage of the Russell Grimwade Building was opened, Max Marginson wrote that 'in the early 1950s eighty students at a time worked in a decrepit laboratory which boasted just one power point! Burettes, pipettes and measuring cylinders comprised the only quantitative apparatus'.[7]

The new building marked a change in the university's architecture and was the first in a series of what Goad and Tibbits have described as 'modernist slab blocks marching north from Grattan Street, each perpendicular to Royal Parade'.[8] It had a steel frame with lightweight floors made of prefabricated metal units. The windows of the north side were framed and hooded with grey, anodised, ribbed aluminium. Construction finally commenced in September 1956 and the first two floors of the building, to be used primarily for teaching, were opened on 16 April 1958.

The Chancellor provided a succinct summary of Russell Grimwade's contribution to the university in his address to the council the day after his death:

> Sir Russell had a direct and forceful approach to all problems coming before the University. In debate he was brief and to the point, lucid in the expression of his views, receptive to the opinions of others and at all times courteous towards those who did not share his opinions. His wise guidance was not confined to the proceedings of Council and its Committees, but was freely sought and willingly given to any Department which sought it, as many did.[9]

Shortly before the opening of the building, the Vice-Chancellor reported that Trikojus had suggested that, in recognition of the generous help given by Lady Grimwade for the completion of the Russell Grimwade School of Biochemistry, the school's library should be named the Mab Grimwade Library. Lady Grimwade accepted the suggestion. Trikojus also suggested that two laboratories in the new school should be named the William J. Young Laboratory and the Ivan Maxwell Laboratory, in honour of his predecessors in the teaching of biochemistry in the university. The proposals were approved.[10]

Mab Grimwade's generosity had indeed been considerable, and the council *Minutes* of 7 July 1958 suggest the frustration the building delays imposed on all involved:

> The Vice Chancellor reported that Lady Grimwade had donated a further amount of £20 000 to assist the completion of the Biochemistry School, on condition that works should be started before the end of 1959 and carried through to its conclusion. In addition to Lady Grimwade's gifts totalling £40 000, £200 000 would be provided as a result of the Murray report, but it is not known when this latter amount will be available.[11]

The second stage of the building, consisting of three additional floors housing the research laboratories and a penthouse, was opened on

12 April 1961. It was completed in 1966 with an advanced laboratory and the large lecture theatre later named after Victor Trikojus. The final cost of the Russell Grimwade School of Biochemistry was about £700 000.

Even new buildings have their deficiencies, and today's occupational health and safety officers would react with the same horror as the recently appointed buildings officer, L.R.D. 'Lardy' Pyke (1912–1987), did when Trikojus wrote to the council in November 1961 asking for an emergency telephone to be installed in the Russell Grimwade Building lift. The matter was referred to the staff architect, with the suggestion that the caretakers should lock the lifts when they went off-duty to prevent people being trapped in them. Pyke had already written to Trikojus on 6 October, saying, 'I recently used one of these after hours when the building was empty and was horrified to discover that the alarm merely rings the bell in the foyer'.[12]

A delay in the opening of the second stage of the building caused several problems, as George Paton, the Vice-Chancellor, noted:

I have been forced to approve the recommendation of Professor Trikojus to delay the beginning of the teaching term for Biochemistry by two weeks, namely, from March 13th to March 27th, so that the Department can move completely from the old building into the new building as from, it is hoped, March 13th.

Apart from the need to make this regrettable decision, I wish to point out that there are other essential University needs which can not be met before Biochemistry has moved out of the old buildings. These are the temporary move of the Department of Geography from sections of the first stage of the new Biochemistry building into old Biochemistry, and also important sections of the Departments of Physiology and Pharmacology.[13]

Despite his evident anger at the delays in the teaching program and frustration at the slow progress of building, Trikojus did have some good words for his colleagues. Writing to the newly appointed Professor of Organic Chemistry, Lloyd Miles Jackman, he noted in December 1961, 'You will find Melbourne a place where members of the Professorial Board frequently differ but very rarely quarrel'.[14]

Even after the department was assembled in one location rather than the five over which it had previously been dispersed, there were other issues. On 27 July 1962 Trikojus wrote to the Deputy Vice-Chancellor:

Yesterday ... I explained that drastic economies would need to be practised in all aspects of the Department's organization, including research and teaching ...

There are several reasons for our present financial position, including the following:

1   We are in a new building with an increased research and teaching staff (one new Senior Lecturer, one Lecturer returned from leave) distributed over five floors, this in itself requiring some duplication of equipment.

2   In connection with our new building we received no grant for equipment (the £16 000 available ... to the building cost having been expended on a lift and other non-biochemical items). Last year after moving in we asked for a special equipment grant of £3390 for our senior Science and Agricultural Science students, as these were being taught in the new building for the first time. This was refused at that stage.

3   In the new building there are additional recurring maintenance costs – cold room servicing, cleaning material, servicing of our minimum pieces of major equipment, obtained with funds from outside services ...

Biochemistry, depending as it does on the techniques of chemistry as well as on so many peculiar to its discipline and developed during its rapid progress, is an expensive subject for undergraduate training ... Many of the chemicals used in a biochemistry course are much more expensive than those used in pure chemistry. At present our grant allows us to spend less than £5 per student on apparatus, chemicals and equipment ...

I wish to apply at this stage for an emergency grant for this year of £1500.[15]

Despite all these problems, the completion of the Russell Grimwade School of Biochemistry Building was a triumph for Victor Trikojus, who had worked hard and effectively over many years to achieve not merely an edifice, but a facility which was to prove fit for purpose for many years. With the opening of the Howard Florey Laboratories of Experimental Physiology in 1963, the Microbiology Building in 1965 and the Medical Centre in 1968, the Dean of Medicine, Sydney Sunderland, was able to report to the University Council that the whole Faculty of Medicine was finally accommodated in the south-western corner of the campus.

Evelyn Mary Jean Parkhill (1930–) was appointed to the department in 1959, the year after the first stage of the Russell Grimwade Building was opened, and retired just ten years before the research staff once again moved to different premises in the Bio21 Institute in 2005. She came to the university to work as a technical assistant to Mary McQuillan with a varied career behind her, having initially trained as a nurse before taking a position in the Daylesford Centenary Woollen Mills in Bentleigh. While the history of this mill forms no part of this story, it is poignant to note that although in 1943 it was offering positions for twenty-five girls or women with free meals and a free medical service, by 1950 the business manager told the Arbitration Court that Australian knitted goods could no longer compete with imports from the United Kingdom, where lower wages were paid. In 1951 the mill stood down 200 employees because of a downturn in sales.[16]

Evelyn Parkhill in the amino acid laboratory

Evelyn Parkhill had already published a paper with Syd Leach as part of the Proceedings of the International Wool Textile Research Conference in 1955.[17] At the university she provided technical assistance to Mary McQuillan in her thyroid investigations and, when McQuillan began concentrating less on research and more on lecturing and departmental administration, Parkhill worked with Barrie Davidson and Robert Augusteyn on amino acids. Her work in this field is acknowledged in a number of scholarly papers.[18] When Augusteyn left to take up his position as director of the National Vision Research Institute, Evelyn Parkhill transferred to work in the student laboratories, which to her great pleasure brought her into daily contact with students and support staff as well her academic colleagues. Although she had come to the university without formal qualifications, during her period in the department she took third-class honours in Chemistry Part I and a pass in pure Mathematics Part I as well as attending lectures in Chemistry Part II and Physiology and Biochemistry Part I.

The department's relocation to the Russell Grimwade Building was contemporaneous with several changes in personnel, beginning with the appointment in 1966 of the first Professor of Biochemistry (Medical). Peter Francis Hall (1924–) took his MBBS (1947) and MD (1956) from the University of Sydney and his other degrees from outside Australia: MRACP (1952) by examination, MRCP (1953) in London and his PhD from the University of Utah in 1961. From 1962 to 1966 he was Assistant Professor of Physiology at the University of Pittsburgh School of Medicine and he was to occupy the chair in Melbourne until 1971 when he left to take up a position at the College of Medicine at the University of California, Irvine. He worked there from July 1971 to April 1980 in the Department of Gynaecology and Obstetrics and the Department of Physiology (formerly known as the

Peter Hall, first Professor of Medical Biochemistry

Curriculum of Functional Correlates-B) and served as Department Chair for Physiology from 1971 to 1977. Peter Hall delivered the 12th Mathison Memorial Lecture in 1971.[19]

Following the retirement of Victor Trikojus, a second professor was appointed in 1968. Simon Joshua (Syd) Leach (1920–2005) BScTech (Manc) PhD, DSc (Leeds) came to Australia from England in 1949 after meeting the new chairman of CSIRO, Ian Clunies Ross (1899–1959), who persuaded him to take a research position in the organisation: he joined the team headed by Gordon Lennox. In England he had spent four years working as a chemical engineer in an oil refinery, after which, at the University of Leeds, he investigated model coenzymes.

At CSIRO, Leach's interest turned to protein chemistry, investigating fundamental aspects of the wool fibre and certain purified proteins. This work, building on the pioneering research of Arthur James Farnworth (1923–2006), was to lead to the development of one of the most significant of all Australian inventions – the SiroSet process which permanently pleats woollen material.[20] The process, widely used in trousers, skirts and girls' school tunics, was a welcome boost to the wool industry, which was encountering strong competition, not only from cotton, but also from new synthetic fabrics which held their shape, especially after washing. The 1964–65 work of Leach and his colleagues provided evidence that the mechanism described by Farnworth and his team in 1957 involved disulphide bond rearrangement by thiol/disulphide interchange.[21] Siroset Close in the Canberra suburb of Dunlop was gazetted in 2003 in honour of this Australian invention.[22]

Syd Leach at work at the CSIRO Parkville laboratory in the early 1960s

Leach spent time in 1970 at the National Institute for Health in Washington DC, working on chain reversals and antigenic determinants in staphylococcal nuclease with the 1972 Nobel Prize winner Christian Anfinsen (1916–1995) and Harold A. Scheraga (1921–) of Cornell, with whom he had collaborated before coming to the department. This work brought to Melbourne the technique for synthesising peptide antigens. The 1973 *Research Report* shows Leach's involvement in two additional collaborative projects, one with Elizabeth Minasian and Leo Edmund Reichert Jr of Emory University on conformational properties of luteinising hormone, and the other involving structural studies on leghaemoglobins with N.A. Nicola and Cyrill Angus Appleby of the CSIRO Division of Plant Industry.[23] Leach's obituary describes him as an exacting but patient teacher, whose laboratory became a centre for collaboration.[24] His belief in the collaborative approach to scientific investigation is illustrated in the paper he wrote for the University of Melbourne Science Students Society on the evolution of proteins, in which he noted his laboratory's collaboration 'with a plant biochemist in Kuala Lumpur, a group in the Anatomy Department at Wayne State University in the U.S. and research workers in the Haematology Department of a large hospital in Melbourne'. He also noted that six of the Nobel Prizes in Physiology and Medicine of the 1970s had been won by chemists and biochemists.[25]

Syd Leach's most lasting memorial is probably the annual Conference on Protein Structure and Function at Lorne. The first, convened in 1976, attracted sixty-five participants from Victorian universities and research institutes. The keynote speaker was Frederick W. McLafferty (1923–) of Cornell University, who pioneered a number of analytical chemistry techniques, including gas and liquid chromatography combined with mass spectrometry. Subsequent speakers who delivered what has since 1986 been known as the Leach Lecture include Elizabeth Blackburn, Adrienne Clarke, David de Kretser and Geoffrey Tregear. The Leach Medal has been awarded since 2004, and Leach himself was awarded Medal of the Order of Australia for services to science, particularly in the field of protein chemistry, in 2000.

When Frederick Darien Collins (1918–1974) came to the department in 1958, he already had a considerable research achievement behind him.[26] Born in New Zealand, Collins took his BSc and MSc (1940) from Victoria University of Wellington and worked as a research officer in the New Zealand Department of Agriculture. He travelled to England in 1946 as a New Zealand Research Scholar in the Department of Biochemistry at the University of Liverpool, from which he took his PhD in 1949. There, as ICI research fellow, he investigated the biochemistry of vision. This work resulted in over twenty papers published in *Nature* and *Biochemical Journal*.[27] His next appointment was to the ANU where he worked as a senior research scholar in the Department of Biochemistry, beginning the investigation of phospholipids which was to engage him for the rest of his life.

Collins came to the Melbourne Department of Biochemistry in 1958 as a research fellow funded by the NHMRC, taking up a senior lecture-ship in 1972. He had been working for some years on phospholipids and publication of the results of his investigations in *Nature* in 1960 'represented a turning point in the importance attributed to these molecules by biological scientists'.[28] Through his work on essential fatty acids, he was able to demonstrate the chemical indicator of the syndrome known as essential fatty acid deficiency and thereby inspire changes in the composition of intravenous food supplements. As well as affecting the nutrition of

Frederick Collins in his laboratory at the Australian National University

diseased adults, these findings influenced the development of improved infant formulas. In the 1970s he investigated the basic biochemistry of essential fatty acids in skeletal muscle phospholipids. In 1977, three years after his death, the results of this research were published in *Australian Journal of Chemistry*.[29]

Collins was active in the establishment of the Australian Biochemical Society (now the Australian Society for Biochemistry and Molecular Biology), serving as its foundation treasurer in 1955 and its second secretary from 1956 to 1961. His family established the Fred Collins Award, which is made to the most outstanding of the society's fellowship applicants and provides $1000 over and above the fellowship travel expenses.

Peter Hall was succeeded in the Chair of Biochemistry (Medical) by Gerhard Hans Schreiber (1932–) DrMed. DozBiochim, who had migrated to Australia from Germany in 1973. Having graduated in physics at the University of Mainz in 1955, Schreiber took his medical degree and postgraduate qualification from the University of Freiburg-im-Breisgau. Before coming to Melbourne, Schreiber had worked at the McArdle Laboratory for Cancer Research in Madison, Wisconsin and Columbia University, New York, followed by seven years in the Department of Biochemistry in Freiburg.

In Melbourne, Gerhard Schreiber was to be recognised as an outstanding teacher and investigator. His most significant discovery was perhaps the finding that transthyretin, one of the major thyroid hormone transport proteins synthesised in the liver, is also synthesised in the choroid plexus of the brain in the region of the blood brain barrier. This discovery created a major new field of research in which Schreiber was recognised internationally as the leader. In addition to revealing its significance in the context of thyroid hormone transport in blood and across the blood brain barrier, Schreiber's structural characterisation of the transthyretin from a broad range of vertebrates identified variations of considerable evolutionary significance in the transthyretin gene between species. On his retirement, the Minute of Appreciation read to the Academic Board noted: 'In addition to his excellent research, Professor Schreiber has made major contributions to the teaching of undergraduate biochemistry and molecular biology in the Medical and Science degree programs of the University, as well as to the training of BSc Honours and PhD students.'[30]

Gerhard Schreiber, second Professor of Medical Biochemistry

Gerhard Schreiber was an active member of the Australian Society of Biochemistry and won its LKB medal in 1983. Ten years later he won recognition of another kind when a clue in the Giant Crossword in *The Age* asked, 'Currently working on blood proteins and their evolution, which German-born scientist was appointed to the chair of biochemistry at Melbourne University in 1973?'[31] The reception staff in the department spent most of the following Monday morning giving callers the answer: 'Schreiber'.

One of the workers in the Schreiber laboratory who went on to a notable career, first within the department as an ARC postdoctoral researcher and subsequently as a ARC research fellow, before leaving for work first in France and then at RMIT University, was Samantha Jane Richardson (1967–). She took BSc (Hons) in 1990 and her PhD on the evolution of extracellular thyroxine-distributor proteins in vertebrates in 1995. The Richardson laboratory focused on the evolution of the thyroid hormone distributor protein transthyretin (TTR), suggesting that the TTR gene had arisen before the divergence of vertebrates from invertebrates. They identified genes coding for TTR-Like Proteins in genomes from all kingdoms and that the protein product was almost identical to that of human TTR; in other words, that the X-ray crystal structure of TTR-Like Protein from Salmonella was superimposable over that of human TTR. Non-vertebrates do not, however, have thyroid hormones, so they identified the function of the TTR-Like Protein from Salmonella as an enzyme involved in uric acid degradation. Thus, a subtle change in a few amino acids led to a drastic change in function. Thyroid hormones are involved in the regulation of growth and development, particularly of the brain. The work of the laboratory revealed that TTR null mice have a developmental delay in thyroid hormone-regulated events, including maturation of the brain.

When her ARC Fellowship ended, Richardson continued her research in the Muséum National d'Histoire Naturelle in Paris, working from 2006 to 2007 with Professor Barbara Demeneix. They discovered that the stem cells of adult TTR null mice had abnormal cycling, similar to hypothyroid wild type mice. In 2007 Richardson moved to RMIT University where she is currently Associate Professor in the School of Medical Sciences and continues her stem cell research in TTR null mice; she is also starting work on TTR amyloid formation.

Among Samantha Richardson's many distinctions are the 2001 Australian Academy of Science Young Researcher Award, and, from the Royal Society of New South Wales, the Edgeworth David Medal for distinguished contributions carried out by a young scientist in Australia. In 2013 she was elected to the Council of the International Federation of Comparative Endocrinological Societies.

One professor was sadly to occupy the Chair in Biochemistry and Molecular Biology for under a year, although he had spent almost his entire working life in the department and made a considerable impact on its research and teaching. Barrie Ernest Davidson (1939–2000) won an Australian Dairy Produce Board Junior Postgraduate Studentship in 1960 and graduated BAgrSc in 1960 and BSc in 1962. He was a demonstrator and then senior demonstrator in the department from 1961 to 1964, when he left to work for four years at the MRC Laboratory of Molecular Biology in Cambridge: he spent the first two years as a CSIRO postdoctoral fellow and the next two years as a medical research fellow. Davidson returned to take up a three-year Queen Elizabeth II Fellowship in 1967 and remained in the department for the rest of his life. He had a profound impact on every aspect of its work. At the end of his fellowship, Davidson became a lecturer. He was a senior lecturer from 1971 to 1986 and a reader from 1986 to 2000, when he was finally promoted to the rank of professor.

Barrie Davidson (*left*) with Richard Pau

Recommending his promotion to senior lecturer a mere year after his appointment, Francis Hird noted that he was:

A very able member of staff. He plays a prominent part in every phase of the Department's activities. He is a responsible man, a capable teacher and very competent in the supervision of postgraduate students in research. He adds a great deal to the Department's reputation, both with undergraduate students, and, at the longer distance, with his research.[32]

The Minute of Appreciation read to the Academic Board after his death recorded that: 'From the very first, Barrie was a dynamic force within the Department, initiating novel areas of research and instituting new lecture and practical courses in modern topics in Biochemistry (and, as soon as the field evolved, Molecular Biology).'[33]

Davidson's work in Cambridge, which was continued on his return to Melbourne with the establishment of a dedicated research group, led to a groundbreaking paper in *Nature* on the amino acid sequence of glyceraldehyde 3-phosphate dehydrogenase from lobster muscle.[34] His later investigations of the transcriptional regulation of gene expression in bacteria, which attracted considerable financial support from industry and scientific research organisations, included research into bacteria and bacterial viruses of significance to the dairy industry.

In 1996, the Cooperative Research Centre for International Food Manufacture and Packaging Science was established with the aim of placing Australia in a good position to win a share of the food market in Asia. Research into bacteriocins (the bacterial peptides produced by lactic acid bacteria) was led by Davidson. Collaborating institutions included Arnott's, Goodman Fielder, various divisions of CSIRO, Pratt Industries and several Australian universities.[35]

Like so many other biochemists before him, Barrie Davidson was a great supporter of the staff club and served as its president from 1988 to 1990.

Unlike Barrie Davidson, William Hugh Sawyer (1940–), who was an active member of the department from 1968 to 2000 and an emeritus professor for two years after that, spent much of his early career outside Australia.[36] Sawyer took his BAgrSc from Melbourne in 1961 and travelled immediately to the University of Minnesota, where, under Robert Jenness (1917–1998), he studied dairy chemistry, investigating why condensed milk tends to gel on storage and finding that this could be prevented by minimising the heat treatment applied during processing. He took his MSc from Minnesota in 1962 and his PhD in 1966 from the ANU, supervised by Laurie Nichol. He then spent a year at the Lister Institute of Preventative Medicine in London working with Michael Creeth (1924–2010), whose own PhD work had confirmed the existence of the special hydrogen bonds which hold the two strands of the DNA molecule together, a finding which proved crucial for the Nobel Prize–winning discovery of the double helix six years later by James Watson and Francis Crick.

Returning to a lecturer's position at Melbourne in 1968, Sawyer investigated plant lectins, notably the way the microheterogeneity affected the self-association and valency of concanavalin A. This work, in collaboration with Donald Winzor, culminated in a monograph on the quantitative characterisation of ligand binding, which was published

William Sawyer (*right*) receiving the 1980 Syme Medal shared with LeRoy
Henderson (*left*) from the Chancellor, R.D. Wright

in 1995.[37] Promoted to senior lecturer in 1973, Sawyer shared the David
Syme Research Prize in 1980 with LeRoy Henderson from Sydney
University's Department of Mechanical Engineering. Sawyer's award
was for 'outstanding contributions to methods for the study of cell
membranes', demonstrating how these methods could be used to investigate cells in normal and diseased states. With his research group, Sawyer
was responsible for the design of an instrument which 'found extensive
application to problems in colloid chemistry, veterinary science, cell
biology, nutritional biochemistry and haematology'.[38]

In 1973, Sawyer began an investigation with Anne McDougall and
Victor Ciesielski of the Centre for the Study of Higher Education into an
area with which he was to become increasingly concerned: comparing the
efficacy of lectures with computer-assisted instruction in physical biochemistry.[39] He was the first staff member at the University of Melbourne
to introduce computer-based exercises into undergraduate science teaching and, in 1991, the same year he was promoted to reader, he appointed
Graham Parslow to pursue this development. Bill Sawyer was appointed

to a Personal Chair in Biochemistry in 1991, a position he occupied until his retirement and acceptance of the position of emeritus professor in 2001. He won an A.G. Whitlam Scholarship in Intellectual Property Law in 1990 and in 1991 became the first University of Melbourne staff member to be awarded a Graduate Diploma in Intellectual Property Law. This in turn led to his membership of the University Patents Committee and Intellectual Property Committee, as well as an appointment as a non-executive director of Rothschild Bioscience Managers Limited, a venture capital trust for the promotion of innovation in science and technology which was bought out in 1996 by GBS Venture Partners. Bill Sawyer was president of the Australian Society for Biophysics from 1986 to 1988 and of the Australian Society for Biochemistry and Molecular Biology from 1990 to 1992. From 1999 to 2001 he was president of the Federation of Asian and Oceanian Biochemists and Biologists, which was founded in 1972 by the biochemical societies of Australia, India and Japan.

During 2001 and 2002 Sawyer was one of the six foundation executive directors of the Australian Research Council and was responsible for assessing all biological science and biotechnology applications. At the time of his appointment, the Australian Government had doubled the funding for the ARC, which made for an extremely busy time making improvements to the assessment system and instituting new programs such as the Federation Fellowships and new Centres of Excellence.

On his return to Melbourne, the Research Office asked Sawyer to suggest measures to improve the research profile of the university. His time at the ARC had made him acutely aware of the difficulties facing early-career researchers and he suggested the establishment of a skills workshop for them. The annual workshop, which Sawyer convened for ten years before standing down in 2013, covers research planning, career planning, communication skills, time management, people management, research ethics, grants and publishing.

Bill Sawyer describes his recreations as swimming, carpentry, viticulture and winemaking – until early 2014 he ran a winery, Wyuna Park, on Victoria's Bellarine Peninsula, which specialised in handcrafted Pinot Noir and Pinot Gris wines. In 2013 applications were called for the inaugural Bill Sawyer PhD Scholarships in Biochemistry and Molecular Biology. Up to five scholarships designed to assist students from interstate or overseas are available. They encompass the areas of molecular cell biology, protein structure and function, biophysics, molecular parasitology, molecular and cellular immunology, inflammation, cell signalling, cancer, neurodegenerative disease, bioinformatics and inherited diseases.

Another long-standing member of the academic staff began his scientific career relatively late, coming to the university under the Commonwealth Reconstruction Training Scheme which was introduced in March 1944 to provide educational and vocational training to those who had served in Australia's armed services during World War II. Robert William Henderson (1916–2013) had enlisted in Port Moresby, New Guinea, and served from 1942 until 1944 in the Australian Army Ordnance Corps with the rank of major. He took his BSc in 1951 and was initially employed as a demonstrator in 1952 before being promoted to senior demonstrator the following year. The letter which Trikojus sent to the university accountant justifying this elevation made a point that would have been valid in the case of many of the new ex-service staff of the period, noting that although he had only recently taken his degree: 'He brings to his teaching a maturity which has been beneficial to the classes with which he has been concerned. He is also carrying on an important research investigation in association with Professor Rawlinson'.[40]

In 1956 Henderson and Rawlinson published the results of this work, and in 1957 Henderson was promoted to lecturer.[41] He was appointed senior lecturer in 1962, a position he occupied for almost two decades. During 1959 and 1960, assisted by the Rockefeller Foundation, Henderson took study leave in Europe and the United States, presenting his PhD on electromotive force and related physico-chemical studies on metalloporphyrin systems. He was to specialise in research on haem and porphyrin compounds and their associated proteins for the rest of his career, publishing four

Robert Henderson, World War II veteran and long-serving lecturer

chapters in books and twenty-one journal articles. His study leave at Philipps-Universität in Marburg was devoted to investigating aspects of mitochondrial electron transport and oxidative phosphorylation.

When he retired in 1982 and took up associate status, he was still engaged in perhaps his most important work, which dealt with the binding of porphyrin derivatives to tumour tissues to selectively sensitise these tissues to destruction by visible light. This technique had been under clinical trial in the treatment of cancer and by 1982 Henderson had identified and synthesised compounds with more favourable characteristics than those presently being tested.

It is fair to say that in the period being discussed in the department, men held all the senior positions. One woman, however, spent over twenty years, from 1970 to 1993, in Biochemistry as a highly respected technical officer. Elizabeth Dockuzian Minasian (1928–) was born in Romania. She had studied pharmacy there and entered the année préparatoire for physics, chemistry and biology at the Sorbonne in France before coming to Australia. From 1962 until 1970 she worked under Pehr Edman (1916–1977), the first director of St Vincent's School of Medical Research, who developed the automated Protein Sequenator. During the eight years she spent there, Minasian was in charge of the isolation of myeloma proteins and macroglobulins from blood serum, reduction-alkylation of the pure proteins and separation of the globulin chains in Edman's work on immunoglobulin primary structure. She provided a vivid account of their collaboration in 2002, in a special feature on Edman in *Australian Biochemist*, including one notable anecdote:

> I remember one day when Pehr Edman came to me and asked first thing in the morning 'Do you know how to knit?' I said immediately, very surprised, 'Yes'. Then he said, 'Leave all your work aside and knit for me an electrical mantle for the bell jar on the Sequenator so that we can raise the temperature in the cup to 50°C and be able to perform the sequence'. I left everything, I went into the city and bought a pair of gloves, white glass wool and a pair of needles. It was quite strange to sit in the lab and knit the bell jar pattern with a little hole in one side to enable observation of the reaction inside. The electric wire was inserted between the two knitted layers and it was used in

the Sequenator until it was replaced by chemicothermo reaction by spraying the bell jar with stannic chloride.[42]

Elizabeth Minasian came to the Biochemistry Department as a technical officer grade 1 to work with Syd Leach. Recommending her for promotion to grade 2, he commented: 'She is highly skilled in the use of complex equipment used to study the physical chemistry of proteins and her talents have been used by members of staff in this and other departments.'[43]

Elizabeth Minasian enjoyed collaborating, in conjunction with Syd Leach, with many national and international scientists. This included some of the first collaborations with scientists from China on the antigenicity studies on peptides synthesised and assayed on resin support, and with scientists from Johns Hopkins University on antiviral activity and conformational characterisation of mouse–human α-interferon hybrids. She published a review with Nicos Nicola on cytokine structures.[44] She also collaborated with Kenneth Sikaris on a secondary predictive conformation on proteins, which was later used in practical classes in the advanced teaching laboratory, and with Herbert Treutlein on a new technique for conformational analysis of peptides and proteins and conformational relationships.[45] She was also a regular participant in the annual Lorne Conference on Protein Structure and Function and co-authored several journal articles with Leach and others, including two on cytokine conformations published in *Journal of Molecular Recognition*.[46]

Elizabeth Minasian, (*left*) and Agnes Henschen (*right*) preparing an amino acid analyser, c. 1967

Victor Trikojus, as we have seen, like many heads of department before and since his time, was obliged to divert much of his attention from teaching and research to departmental and university administration. This necessity had plagued Medical staff from the establishment of the Medical School, with Professor George Halford widely regarded as not having fulfilled his research potential because of other demands on his time. When Lloyd Finch reluctantly succeeded Mauritzen in 1980 for the first of his two terms as Head of Department, he saw his main task as building up its graduate school, beginning by rebuilding his own research unit as an example to the department. He had a considerable research and teaching career on which to build and recalls his first term as Head of Department as being principally devoted to this. Finch was succeeded by Sawyer, in turn succeeding him for a second term from 1983 to 1985.

Lloyd Ross Finch (1927–) took his BSc (Hons) from the University of Western Australia in 1950. He worked as a research officer at CSIRO for two years before joining the department in 1952 as a part-time demonstrator and research assistant to Legge in his work on the purification of secretin and formation of induced (adaptive) enzymes in bacteria. From 1957 to 1958 he worked on a CSIRO Overseas Studentship at the Medical Research Council Experimental Radiopathology Research Unit of Hammersmith Hospital, taking his PhD and lectureship in the department in 1959. The teaching load was considerable. At the beginning of the 1963 academic year, Trikojus wrote to the Deputy Vice-Chancellor:

> Previously the teaching load for the nearly 180 students in Medicine Division IIA had been undertaken by Dr Finch, Mr Marginson and Dr Stone. However, I have now transferred Dr Stone to assist Dr Hird in the Agriculture class (fourth year). This leaves two members of the academic staff, both with Lecturer status, Dr Finch and Mr Marginson, as responsible for one of our largest classes. In addition, Dr Finch contributes an important group of lectures to the final Science year, as he is the acknowledged expert in Melbourne on nucleic acids, particularly in their relationship to protein biosynthesis. Hence, he was selected as the main contributor to the recently conducted Cell Biology Symposium, held at the University of Melbourne, with participants from other parts of Australia.[47]

Lloyd Finch at the
Congress of the
International Organisation
for Mycoplasmology,
Istanbul, 1990,
talking with Professors
Janet Robertson and
Gerry Stemke

Finch was promoted to senior lecturer the following year and in 1965–66, on a US Public Health Service International Postdoctoral Fellowship, he worked at MIT under Boris Magasanik (1919–2013). Like so many others, he was impressed not only by the ready availability of funds and facilities for research in the United States, but also by differences in undergraduate teaching, reporting to the University Council in 1967:

> I was given a laboratory and invited to procure whatever I needed to expedite the research project that I was undertaking – on nucleotide concentrations in relation to nucleic acid synthesis in bacteria.
>
> Initially the excellence and abundance of the research facilities were somewhat overwhelming. Because of the high level of financial support for biological research in the United States, much can be bought ready-made for the researcher so that he is freed to concentrate directly on attacking the main aspects of his problem rather than being forced to devote a major part of his effort just to providing means. This leads to a directness of approach, a professionalism in research, whereby one can expect, and be expected, to put one's ideas to the test of experiment. There is considerable incentive, and also obligation, in knowing that the potentialities of one's work are sufficiently thought of for it to be supported fully.
>
> Liberal finance, however, was only one of three advantages that I was strongly aware of to researchers in the U.S.A., or at least in the Boston area. A second was the rapid availability of new knowledge

which flowed in from many sources other than the scientific journals
… The third advantage was the ready access to expert 'know-how'
in many aspects of the discipline. An unfamiliar technique could
be readily undertaken because one could go directly for advice, and
frequently facilities, to someone who was a leading exponent of it …

Another factor contributing to the flexibility that I observed in
American graduate students and post-doctoral workers may lie in the
less specialised nature of their undergraduate training and the multidisciplinary approach to training students in biology, at least in many
departments.[48]

Altogether, Lloyd Finch spent nine years as head or deputy head and
in 1988, with only four years till his own retirement, he refused to be
nominated for deputy chair, asserting that he felt that the department
would be better served by someone with a longer period of future commitment and a fresher approach. His research interest in the meantime
had switched from metabolism to molecular biology. A paper on quantitative extraction and estimation of intracellular nucleoside triphosphates
of *Escherichia coli* is among his most cited and, reflecting his change of
research interest, the year of his retirement saw the publication of a book
he coedited on the molecular biology and pathogenesis of mycoplasmas.[49]

The next professor appointed to the department was to oversee a
very new development of its research effort with the establishment of the
Bio21 Molecular Science and Biotechnology Institute and the eventual
abandonment of the very building Russell Grimwade and Victor Trikojus
had fought so long and hard to gain. One of Lloyd Finch's principal concerns, however, had been that while the department had long been seen
as a valuable place in which to gain basic training, students wishing to
pursue a career in biochemistry were forced to look interstate or overseas
for initial employment or promotion.

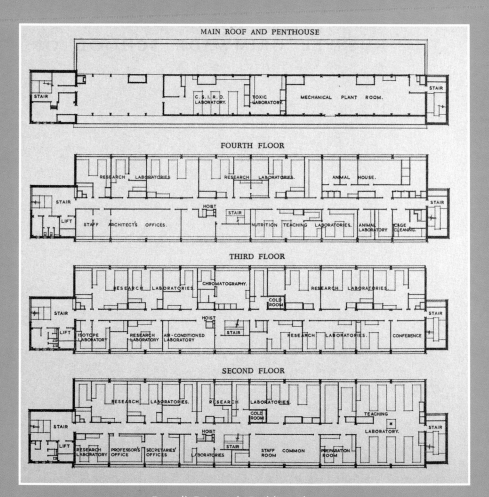

Russell Grimwade Building plan

CHAPTER 6

The Russell Grimwade
School of Biochemistry

THE OPENING AND occupation of the Russell Grimwade School of Biochemistry and Molecular Biology Building were eagerly anticipated and much celebrated, but the structure lasted just fifty years. Opened in 1958, it was demolished in 2007, when the department was once again, this time with forethought and deliberation, divided. Still, the Russell Grimwade Building was in many ways a leader in its field, the first in a series of buildings of similar appearance and construction. It predated the first stage of the Howard Florey Institute for Experimental Physiology, which was opened in 1962, and the Microbiology and Immunology Building, the first stage of which was constructed between 1962 and 1964. It also predated by almost a decade the triradiate Medical Building on Grattan Street, which opened in 1969. Discussions about the design, cost and building program for the new school edifice preoccupied the University Council and its Buildings Committee from the mid-1950s onwards.

In February 1954 Walter McCutcheon submitted preliminary estimates for four options. The Buildings Committee recommended the second-cheapest at an estimated cost of £199 000, the option also favoured by the architects.

> This scheme provides for the erection of the whole framework of the building for the basement, ground, first and second floors in nine bays. The building will be roofed and from the outside will appear complete, but only five bays on the ground floor and two in the basement will be completed internally.
>
> Professor Trikojus pointed out that this scheme will house only part of his department, and that it will be necessary to complete the interior of the rest of the building as soon as possible to avoid having his department in two buildings some distance apart.[1]

The reasons provided for favouring this proposal made clear some of the complexities and constraints:

1   Building involves many trades each following the other in a sequence which allows for little variation. The initial operations connected with the erection of the building are related. If a building has to be built in stages, it is generally easier, cheaper and more efficient to erect the structure first, and furnish it at a later date: than to erect and finish part of the building first and to erect and finish the remaining portion later on.

2   The erecting trades – excavators, concreters, steel riggers, bricklayers, etc. – require large areas around the building in which to operate, and are dirty, noisy and generally more offensive than those connected with the finishing of the building.

3   Completing the exterior of the building at the one time will ensure the uniformity of the external finishes, and will obviate the construction and expense of temporary ends and escape stairs. It also means that the external appearance of the building in the first instance is complete and satisfying.

4   The building is being designed so that a high degree of flexibility is possible in the layout of each floor. The building may be finished therefore as occasion requires. As only finishing trades are involved, the occupied portions of the building will not be

seriously interfered with. On the other hand, if structural work were involved, occupation of the finished portions would be at least, difficult.

5 Service mains will be designed to enter the building at the most appropriate points, and must be of sufficient size to take the ultimate load of the finished building. It is expected that the best layout of services, and also the cheaper, will result from erecting the whole carcass of the building at one time.[2]

Discussions, proposals and counterproposals continued throughout 1954 while at the same time the university was still obliged to spend money refurbishing the department's existing quarters. In February 1955 the Buildings Committee recorded that matters dealt with by the buildings officer included £10 for painting the dado in Mrs Winikoff's room, £65 for a partition in the basement and £80 to replace a siphon pump with an electric one. In May 1955 the department spent £75 on cupboards and shelving, £72 on wooden stools, £26 on an animal operating table and £40 on a mail box, blackboard and doors for shelving.

Early in 1958, the university's Maintenance Department took possession of new quarters at 804-816 Swanston Street. The Faculty of Medicine took over the old quarters and Trikojus objected to dogs being housed there, asking for space to be made available to Biochemistry. The Buildings Committee left decisions on the allocation of space to the Vice-Chancellor but asserted that the animals should indeed be accommodated elsewhere.

The Russell Grimwade Building was only one of several projects in the university pipeline at the time, and some of these had a direct bearing on its development. For example, the Medical School, which was not to get its own new building until 1969, had strong opinions about the Biochemistry one. Construction of the Baillieu Library was underway at the same time as that of the School of Biochemistry and the cost of the necessary additional sewer line running from the Union Building to Grattan Street was to be shared, as was the cost of a stormwater drain.

Throughout 1957 and 1958, additional expenses came to light. In April 1957 the architects pointed out that the existing contract did not allow for the necessary excavation, the building of a retaining wall and steps, and the proper siting of drains to the north of the building. They recommended that this work be undertaken in such a manner that when the building was completed, none of the work would need to be lost. They estimated the cost of this at £4000. The roadway between Biochemistry and the Baillieu Library was the subject of almost endless discussion relating to its height, width, distance from Sydney Road, trees along its border and so on.

In November 1957, the Finance Committee received an application from Trikojus for £14 500 for equipment for the new building, the bulk of which would be needed because the school was obliged to operate in two locations. The committee recommended an immediate allocation of £5000 so that advance orders could be placed, while noting that no allowance had been made for this in the 1958 estimates presented at the previous meeting. On 9 December it recommended an additional £5000 for equipment, of which half would be recoverable through the sale of Students' Sets. With the £5000 already allocated, Trikojus would get two-thirds of the amount he had requested.

Trikojus had not for one moment allowed the university to think that opening the first stage of the department's new quarters would see an end to his campaign for the completed building. In November 1957, he made a 'Statement concerning the Russell Grimwade School of Biochemistry':

1 It is important to realise that the Russell Grimwade School of Biochemistry is not a new project. Sir Russell made his gift of £50 000 in 1944. (Messrs Nicholas Pty Ltd added to this sum by a grant of £20 000 in 1946.) Recently (1957) Lady Grimwade contributed a further £20 000 towards the School. Thus, with interest, over £100 000 is available from benefactions. With the background of Sir Russell's gift, the University hoped to begin the construction of a modern school of biochemistry soon after the end of the war.

2 This was 3 years ago. With rising building costs, coupled with the restricted nature of University finance, it is even now only possible to complete two-fifths of the building agreed as essential for the subject of biochemistry – a scientific discipline which has undergone tremendous advances during the past decade.

3   The present section of the building is expected to be finished by the end of November, 1957 (providing mainly teaching laboratories).

4   If there can be no early planning for the completion of the Russell Grimwade School, there will develop a chaotic situation in that teaching laboratories and most senior staff (with their research units), including the Professor, will be separated by some 550 yards (on the opposite sides of the University campus) …

5   The Department of Biochemistry is at the moment housed in three separate buildings – two of them part of Old Chemistry – while the third contains the main teaching laboratory, which was formerly the dissecting room of Old Anatomy. The Department's workshop is on the other side of the University (housed in the School of Agriculture). Under these conditions, the Department caters for students in four different Faculties – Medicine, Science, Agricultural Science and Dentistry – and endeavours to carry out a co-ordinated research programme.[3]

In April 1958 the Buildings Committee endorsed the Vice-Chancellor's approval of £864 for bookshelves for what was to be named the Mab Grimwade Library. Following protests from Trikojus about damage done to the south-facing windows because of ball games played on the hockey ground, it also decreed that a 10-foot cyclone fence with 15-foot posts would be erected for 20 feet on either side of the centre line behind the northern goal post. Although hockey was later no longer played there, this would not be the last threat to the Russell Grimwade Building's windows from sporting activity. In October 1959 the Buildings Committee rejected a request from the secretary of the Recreation Grounds Committee for hammer throwing to be permitted in the same area.

In July 1958, just a few months after the opening of the first stage on 16 April, Lady Grimwade took a decisive hand in expediting the completion of the building

Mab Grimwade, philanthropist

her husband had not lived to see: she donated another £20000 on condition that the work should commence before the end of 1959. This certainly got things moving, and on 18 August the Buildings Committee received a report from a subcommittee consisting of Professor Amies, Buildings Committee chairman E.T. Beruldsen and Professor E.S. Hills, which reported:

(a) Completion of the Building

The subcommittee was convinced that three additional floors would be needed. It noted that the number of undergraduates in the Biochemistry school was 550 this year. The sub-committee considered that the completion of the Russell Grimwade School would need to be considered relative to the claims of the new Library, the North Building and the Hydraulics Building, to which the University was already committed. The sub-committee was assured by Professor Trikojus that:

1 If the shell and the third floor only were completed, the red brick building at the N-E corner could be evacuated.
2 If the shell, third and fourth floors were completed, the red brick building and 60–75% of the remainder could be evacuated.
3 If all three floors were completed, the whole of the present building would be evacuated.
4 It would be possible to omit certain of the partitions and fittings on the upper floors so that the three floors could be completed for an additional £300000. An expenditure of this amount would enable the complete evacuation of all the present buildings in Tin Alley.

The Buildings Committee in view of the above recommend:

(i)   The Biochemistry Department requires three additional floors and the cost should not exceed £300000.
(ii) If the sum of £230000 is available, the shell, the third floor and as much as possible of the fourth floor could be completed.
(iii) Residential quarters for a caretaker should not be provided at present.

(b) Cost of Lower Floors

A report from the architect indicated that the cost would be £20376 more than the contract price. The Committee viewed with alarm that £7000 of this had been incurred without official approval by the University. The Buildings Officer was instructed not to approve of any further 'extras'.[4]

In September Lady Grimwade was informed that construction would 'probably' begin before the end of 1959. The architects were told in no uncertain terms that £300000 would be available then and that this was to include building, fittings, furniture and all fees. They were instructed to work with the staff architect, Trikojus and the buildings officer to ensure that the building 'should be completed ready for occupation with the money available'.

Decisions had to be taken on the relative priority of the nutrition laboratory and completion of the junior teaching laboratory. Various economies were approved, mainly in materials – granolithic finish instead of terrazzo; there was to be no mechanical ventilation except for the animal house, although the ductwork was included; laboratory benches would be lower in quality than those in the first part of the building; there would be no lift, although a cathead would be placed above the western stairs for hoisting goods; there would be no finishes in the undeveloped areas.

When the final stage of the building was opened in 1961, Victor Trikojus was justly proud of his great achievement. In recognition of its national and international significance, *Nature* published a detailed description of the 'completed building, of about 60000 sq. ft. [which] is by far the largest in Australia devoted to biochemical science'.[5] The article describes each floor of the completed structure, noting: 'A north extension of three floors at one end houses student facilities, an electrical sub-station, the main entrance and foyer, a lecture theatre and departmental library'.[6]

The ground floor of the main structures accommodated a mechanical plant section, a large-scale laboratory, an area for physical biochemistry and the stores. It was connected to the external bulk solvent store. Three teaching laboratories were housed on the first and second floors and the rest of the building was given over to offices and laboratories for research, with cold rooms on each floor. The fourth floor accommodated the Nicholas Nutrition Laboratories and the animal house. The Sugar

Research Unit of CSIRO was temporarily housed in a general laboratory section on the top floor, together with a second mechanical plant room and a toxic laboratory. The floor plan of each laboratory was intended to ensure maximum flexibility, with benches and partitions able to be moved as required. The article concluded optimistically: 'From small beginnings the school now has a full-time teaching, research and technical staff of about seventy. A broad pattern of research is in progress, and all members of the staff look forward to a period of vigorous development in their new environment'.[7]

It had been a long time coming. The Russell Grimwade School of Biochemistry was the first new building for a technical faculty of any magnitude constructed at the University of Melbourne since the opening of the new Chemistry School in Masson Road in 1938. At the opening of the first stage of the building in April 1958, Trikojus had made his frustrations clear to an audience which, sadly, did not include the Victorian Premier:

And now, Mr Chancellor, there has been completed the first stage of a building which we trust has been designed efficiently and with a degree of elegance, but without extravagance, and which will stand up to critical appraisal.

It is perhaps pertinent to ask how long we must wait before the Russell Grimwade School is completed. The present stage provides mainly for students. The staff, teaching and research is almost entirely still quartered in the old buildings some 600 yards away on the other side of the campus. It is already becoming patent who are the biochemical staff in this University. They are developing certain peculiarities – heads thrust well forward, arms freely swinging and shoes worn to the uppers.[8]

He also provided a brief account of the peregrinations of the department before 1958, recalling that in 1943, the main laboratory (which had originally served as the Anatomy dissecting room) was above the Pathology laboratory. Another laboratory was in the Physiology Department and there were 'a few scattered rooms and offices'. The department subsequently spread into Old Chemistry, some distance from the main

Russell Grimwade Building

laboratory. As Trikojus himself noted, the university had been obliged to develop a remarkable ability to adapt any roofed area to a variety of purposes.

The architects, Bates, Smart & McCutcheon, had had a hard task-master. Trikojus commented that he doubted that any architectural firm had ever drawn more miles of lines for a building of comparable size. As his biographers in *Historical Records of Australian Science* recorded:

> The result of these frustrations was not wholly bad. Few laboratories have been designed and redesigned so many times, or subjected to so much criticism by its potential users. Little of this criticism was accepted by the arch-designer, who spent seemingly endless hours, day and night, annotating the architects' dyelines and meticulously drafting his own amendments. The most serious consequence of the delays was that the building had to be constructed in two stages, so increasing the final cost and further postponing the time when the staff would all be under the same roof.[9]

About a year after the opening of the first stage of the Russell Grimwade School of Biochemistry Building, Trikojus was still express-ing his dissatisfaction at the need to wait for its completion. In July 1959, after specifying a large increase in staff that would be required by 1967, he strongly recommended the appointment of a staff engineer to report at regular intervals on the state of the university's mechanical plant. In fact, the first person to bear this title was Arthur Kinsman (1913–1994), who was appointed in 1966; it was only in 1963 that a mechanical tech-nical services officer, directly responsible to the buildings officer for the proper maintenance of mechanical services within the university, had been appointed. Before that, such maintenance had been undertaken by individual departments.[10]

Overcrowding and lack of funding were serious issues in 1959, with Trikojus noting:

> At present students in all Faculties are taken in the Maxwell Labora-tory, with the exception of senior students in Science and Agricultural Science whose classes are held in Old Biochemistry. In classes over

80 some students are accommodated in the Young Laboratory but in 1961 this will be needed for senior Science and Agricultural Science students. With Maxwell Laboratory used exclusively for Medical Students a maximum of 240 could be accommodated ...

Although tenders have not yet been called for the upper floors of the new School, calculations have been closely made and it is not anticipated that the Junior Laboratory can be added within a contract price of £300 000. Other deletions have been necessary – e.g. the laboratory for the training of students taking the Nutrition course. Moreover, it will be essential to complete at least parts of other areas left underdeveloped if new staff is to be accommodated. An additional amount of about £30 000 would provide for the various essential items which we have been forced to exclude at this stage.[11]

There were also minor problems associated with the finish of the laboratories. Replying to a query from the government analyst in Hobart about epoxy coating of benches in the Russell Grimwade Building, Trikojus noted that:

Unfortunately, the material was not supplied by the contractors strictly in accord with the suppliers' instructions. We have found those benches where the resin was well applied to be fairly resistant to organic solvents including $CHCl_3$, CCl_4, BuOH, ether, EtOH. Acetone attacks slightly; pyridine and phenol attack readily. Dilute acids have no effect. Concentrated acids do attack the surface, particularly oxidising acids such as H_2SO_4 and HNO_3. The resistance to alkalis is good. With care to avoid too much reflected heat, the heat resistance is good.[12]

Mab Grimwade continued to be a constant and canny source of support after the opening of the first stage of the Russell Grimwade Building. In 1960, she donated £10 000 towards items as various as a visiting research fellow's laboratory, an additional washing-up unit in the Young Laboratory, and a glassblowing room.

Planning for the second stage of the building was complex, with various parts of the university and the scientific community more generally competing for space in the new facility. In June 1960 Trikojus reported to the University Council that Ian Wark of CSIRO had approached the university with a request for 1000 square feet to be made available on the third floor for a Sugar Research Laboratory. Trikojus noted that there was no money to finish the area in question and he would recommend approval only if CSIRO or CSR paid for the work and vacated the space in December 1964. Disquiet was expressed in the council, especially by Arthur Amies, at the idea of allowing outside organisations to use university space if another department could use it.

This stance was somewhat undermined by a letter from Alan Buchanan of Chemistry, to the effect that it would not be practicable for his department to use the 1000 square feet available, but that Chemistry would welcome the presence of the CSIRO Sugar Research Laboratory. Despite assertions from both the Vice-Chancellor and the Chairman that the university would be open to severe criticism if an outside body were allowed space when it was at such a premium, and expression of the view that the space should be used for administrative offices and occupied by the staff architect, it was resolved that if CSIRO believed it feasible, the university would make space on the roof available for the Sugar Research Laboratory, provided CSIRO paid for fitting it up.

CSIRO attached considerable importance to this development, announcing its establishment in *Nature* at the end of July 1960, noting:

> The decision to establish the Laboratory has come after considerable discussion with representatives of the sugar industry, which is anxious to find new uses for the present surplus in Australian production. The present surplus corresponds to about a million tons of sugar cane per year. Australia's biggest sugar producer, the Colonial Sugar Refining Co. Ltd, has agreed to contribute £2500 per annum towards the running costs of the Laboratory.[13]

On 15 August 1960 the Buildings Committee heard that Bates, Smart & McCutcheon estimated that accommodation for the staff architect would cost £4450 and the Sugar Research Laboratory would require £8850. Roofing for a penthouse would cost £1750. This was the area in which it was proposed to accommodate Jack Clark, the caretaker of the

Old Arts Building. He was currently living in 'Vectis', the fifth house in Professors' Row, which was in turn to be used to accommodate part of the Department of Education. CSIRO agreed to contribute £9260 to the Sugar Research Laboratory – £3000 on completion and £6260 after 1 July 1960 – provided it was guaranteed rent-free occupancy for five years. CSIRO vacated the laboratory at the end of June 1965, leaving it to be occupied by Frederick Collins and his research unit.

The controversy within the university over the accommodation of the Sugar Research Laboratory contrasts starkly with the spirit that inspired the creation of the Bio21 Institute some forty years later. Bio21 specifically fosters industry relationships and accommodates a number of industry groups, including the CSL Limited R&D group, which is focused on therapeutic discovery; BioScreen Medical, which is focused on microbial diagnostics; Sienna Cancer Diagnostics, which investigates molecular biomarkers for specific tumours; and the Prana Biotechnology Limited Chemistry R&D Laboratory, which researches disease-modifying therapeutics for the treatment of common neurological disorders, particularly Alzheimer's, Parkinson's and Huntington's disease.

The final stage of the Russell Grimwade School of Biochemistry Building was opened on 12 April 1961 by John Eccles, president of the Australian Academy of Science, foundation Professor of Physiology at the John Curtin School of Medical Research, Nobel laureate and Bio-chemistry alumnus. Trikojus had taken particular care to ensure the success of the event. On 28 February 1961 he had written a personal letter to the manager of the State Viticultural Station at Rutherglen:

> Would it be possible to obtain five gallons of your dry flor sherry before April 12th? On this date we are opening the Russell Grimwade School of Biochemistry and there will be refreshments afterwards. As one who is pleased with your product I would like to have this served on this occasion.[14]

Teetotallers were not forgotten, and on 1 May 1961 Trikojus sent a letter of thanks to A.L. Lazer Jnr for pineapple juice donated: 'It was very pleasant to have such a first-class product as part of the refreshments'. The staff on duty were not forgotten either, and a month after the opening

Trikojus submitted a bill to the Vice-Chancellor noting that he had made some payments from his own pocket:

> You will see that there is an item for the Maintenance Department cleaning staff. May I assure you that, without their most efficient assistance on the Friday night, the whole of Saturday and the Tuesday night before the Opening, we would not have had the building clean for the ceremony. I was very impressed indeed with the way they worked and I thought a small token of appreciation was not out of place, as they did not share in refreshments at the Opening.[15]

Eccles had declared himself very happy to be asked to perform the opening ceremony, writing to the Vice-Chancellor:

> Dear George
>
> Thank you for your letter of 31st January. I feel greatly honoured by the invitation to open the Biochemistry Building at Melbourne University on Wednesday, 12 April. It will certainly bring back memories when I was struggling with biochemistry under the auspices of Bill Young. In fact, it will be a few days after the 40th anniversary of my appearance in the biochemistry class of the University of Melbourne. That looks as if it might be a good remark for the opening speech![16]

Despite representing a long-awaited triumph for Trikojus, the Russell Grimwade Building was not without its problems, even before his retirement. In July 1962, he wrote, as we have seen, to the Deputy Vice-Chancellor setting out various problems associated with the new building and requesting an emergency grant of £1500.[17]

In February 1965, as well as asking for an allocation of £60 000 in the 1970–1972 triennium for two additional floors each of 3000 square feet, Trikojus wrote:

> I also wish to suggest some provision for sound-proofing (which may have to include air-conditioning) of the vocal and instrumental practising rooms of the Conservatorium which appear to be exclusively on the south side of that building and which face Biochemistry. It seems

quite unfair that an academic department such as Biochemistry has to be subjected to this disturbance which often goes on from 9 a.m. to late in the evenings. The members of the Conservatorium staff are, I understand, adequately protected, but this does not apply to the staff of Biochemistry.[18]

The following year, he asked the Vice-Principal, Ray Marginson, to find alternative accommodation for the staff architect as the space his office occupied was required for new academic staff. The architect was due to move to the Raymond Priestley Building when it opened in 1967 and it was suggested that he be accommodated in the Microbiology Building until then.

Despite his comments to the government analyst quoted above, Trikojus had been proud of the internal architectural and engineering features of the building. In his paper in *Nature* he noted that:

The pressure services – compressed air, gas, hot and cold water, filtered and demineralised water, and the heating system – are run in horizontal ducts below the window sills and connect to distribution ducts along the centres of the laboratory benches. The fitments, which have been designed in standard units to reduce manufacturing costs, are protected with a plastic acid-resistant finish.

The drainage system is run in corrosion-resisting polyvinyl chloride piping through polythene traps. Wastes are concealed in ceiling ducts and connected to a series of vertical ducts spaced along the centre of the building; thus all pipes and wastes are readily accessible for maintenance. Various parts of the building are mechanically ventilated, including the larger teaching laboratories, by connexion to a system of positive pressure. Plastic tube vents from the fume cupboards rise in the vertical ducts and terminate with individual exhaust fans grouped at 40-ft intervals above the penthouse-level. Automatic fire alarms and extinguishers give protection against the spread of fire and a warning system indicates any failure in the mechanical plant. A passenger lift located in the eastern tower and an electrical hoist from the stores in the centre of the building provide additional services.[19]

In 1966 the final serious alterations were made to the building, with the addition on the northern side of what was named the V. M. Trikojus Theatre in 1982, with an advanced laboratory on two levels and seminar rooms.

Despite this, the Russell Grimwade School of Biochemistry Building was fated to disappear half a century after its first opening ceremony, and all that is left of it now are the various plaques that decorated parts of the original structure. Humphreys suggests some of the reasons for its demolition:

> The north-facing rooms were inadequately protected from sun, the enclosure of steel beams in fire-resistant asbestos was in hindsight a disaster, and there was a superfluity of sinks as more sophisticated products such as ADP were bought off the shelf rather than prepared in the laboratory.[20]

The decision to remove the Russell Grimwade Building entirely was not taken either lightly or quickly: in fact, initially the university opted to renovate the ageing structure from the inside. Once the research staff had moved to the new Bio21 premises, leaving the teaching staff, lecture rooms and laboratories on the lower floors, the upper storeys were intended to be renovated to accommodate Federation Fellow Professor Ary Hoffmann and his team. This proved impossible for a number of reasons, one of which is evident from the allusion to asbestos.

This problem took some time to reveal itself, but twenty years after the opening, in 1987, removal of asbestos began with the Biochemistry staff still located on the lower floors. The upper section of the building was wrapped up and isolated from the remaining floors as the asbestos was removed.

This was not a trouble-free procedure, as exchanges between Biochemistry staff and the Department of Property and Buildings, the General Staff Association and the university's Occupational Health and Safety personnel make clear. In 1988 asbestos was removed from the underside of the slab above the central corridor of the building and from the walls of the six internal service ducts on the first floor; removal of the asbestos from the ground floor took place over January 1990. Between

1987 and 1990 considerable tension was demonstrated as staff repeatedly voiced concerns over the removal of asbestos occurring close to areas in which they and their students were working, and complained during the first two years of the program of inadequate information before removal took place. Part of the problem was undoubtedly due to conflicting views on the danger posed by airborne asbestos fibres. As late as February 1989, the Deputy-Vice Principal (Property) asserted in a letter to the Head of Department that:

> Fortunately the only proven health hazard is to those workers who have had long exposure to asbestos, such as mine workers and perhaps some tradesmen working indoors. Asbestos has been used widely throughout the community and a consistent link between asbestosis and those who have had occasional exposure to asbestos has not been established.
>
> Throughout the community there is a diversity of opinion so I can only be guided by the views of experts who have studied the matter and have formulated the asbestos removal regulations.
>
> This department has always complied with the very strict requirements laid down by the Department of Labour. That Department has to approve the set-up and work procedures and also must be kept continually informed of the results of fibre monitoring at the conclusion of each work shift. Furthermore the whole work area must be kept under negative pressure so that any air flow through microscopic apertures is from outside to inside.[21]

It was not until December 1989, however, that the issue was finally resolved and the Head of Department was able to assure the staff that he was 'satisfied that the operation will be conducted without risk to the health of members of the Department and with the minimum of disruption'. Specifically, he assured them:

> Important considerations have been the reliability of the proposed method for removal and the effectiveness of the monitoring and clean-up procedures. In our case, the already tested and proven operation is to be based on that used at the Royal Children's Hospital. This will be

carried out by the same contractor at considerable extra cost to contractors. I am satisfied that, with the experience of the contractors and
assurances given to me by the university with respect to monitoring
… there is no need for undue concern …

'I'his issue has been a difficult one for all of us but I am now
confident that we have a satisfactory solution.[22]

By the 1980s many other attributes of the building, such as the
underfloor heating, no longer functioned efficiently. More significant to
the university's planning, however, was the actual footprint of the fifty-
year-old structure. The Russell Grimwade Building had been carefully
landscaped with internal courtyards and other features that meant that it
covered an uneconomically large area of land in a small campus in which
every centimetre counted. The Melbourne Brain Centre in the Kenneth
Myer Building is a much larger structure, accommodating laboratories
and facilities that are far superior than would have been possible in the
Russell Grimwade Building.

The decision to demolish the building was not taken quickly, however,
and at times the university seems to have been pulling in two directions
at once. As late as 12 April 2006, the University Infrastructure Committee
reported that:

Federation Fellow Professor Ary Hoffmann and 40 staff and researchers were to be provided with research laboratories and facilities on
Level 3 of the Biochemistry Building. Level 4 of the building was to be
refurbished as research laboratories for Genetics and Zoology.

At Capital Projects Committee on 25 November 2005, it was agreed
that the works in the Biochemistry Building would be cancelled and
Professor Hoffmann's research laboratories would be located in the
David Penington Bio21 Institute Building along with the research
facilities for Federation Fellow Professor Paul Mulvaney.

The decision was taken in part because hazardous materials
removal and demolition works in the Biochemistry Building had
revealed a poor quality building fabric and infrastructure. The building site is also being assessed for development, requiring demolition
of the Biochemistry Building.

Courtyard of the Russell Grimwade Building

On the basis of the above the Committee agreed that all BR works for the original Biochemistry project should be finalized.[23]

On 15 November 2007, the University Infrastructure Committee reported that:

The project involves the complete demolition of the existing Bio-chemistry Building to make way for a three-level underground car park & new Neuroscience Building. A contract has recently been let with the contractor starting on site this week. Site establishment is being carried out at present. The anticipated completion date is mid March 2008.[24]

Despite the apparent contradictions inherent in trying to renovate the Russell Grimwade Building at the same time as its demolition became more and more obviously inevitable, few could fail to sympathise with research fellow Terry Mulhern as he described the state of the building in general and his laboratory in 1999 as 'noisy, draughty, dusty and prone to floods from frequently bursting pipes'.[25]

Knowing what to do with the staff who remained in the building and their associated functions was another problem. It is by no means certain that when the decision to demolish the building was taken, all the decision-makers were fully aware of how much space would be required elsewhere to accommodate the large laboratories and lecture theatres required for the department's undergraduate classes. Paul Gleeson recalls walking round the campus with Deputy Vice-Principal (Property and Buildings) Douglas Daines, inspecting various unsuitable venues, among them the derelict and deserted Dental Hospital on Royal Parade.[26] Gleeson attached considerable importance to situating part of the department in the triradiate Medical Building, and space was ultimately made available there on two floors. In 2014 a new Cell Signalling Centre was also under development within the Medical Building.

The move itself to Bio21 went – perhaps surprisingly considering the amount of highly valuable and fragile material to be moved – both smoothly and quickly.

During November 2004 David Penington, a former vice-chancellor of the University of Melbourne, was commissioned to review the Bio21 Molecular Science and Biotechnology Institute. His report was approved by the University Council in 2005. As well as providing a succinct history of the Bio21 Institute, it highlighted specific areas of concern, relating both to accommodation and the terms of employment of Biochemistry staff. Penington noted that:

> Some academics came to believe that the Institute was being developed specifically to relieve the University from obligations to improve accommodation for these two departments, but their accommodation expectations far exceeded what could be provided in the Institute. In January 2001, the Vice-Chancellor limited the construction program to $95m and the projected commitment of space for researchers from the respective departments was reduced. Running costs were to be derived from research grants, contributions from participating departments/faculties and to be enhanced subsequently from external sources.[27]

The Penington Review also reported disagreements over whether the heads of the departments of Chemistry, Biochemistry, and Molecular Biology and Genetics should have a formal executive role within the institute, and noted:

> A further example of this tension was negotiation over management of the Institute's chemicals store. It was proposed that this be staffed and managed by the two major departments rather than being staffed and managed by the Institute as the central store for chemical supplies, liquid nitrogen, solvents etc. for the Institute as a whole. Liaison with the department store managers, with effective rationalisation, will clearly be necessary in serving members of their departments in the Institute.[28]

Academic staff were also still waiting for letters offering accommodation within the institute as well as details of any conditions attaching to such offers. The review also noted that:

The previously proposed arrangement of joint appointments of 80% to a Department and 20% to the Institute for teaching and research academic staff, and of 50:50 appointments for full-time research appointments, has created many difficulties. Staff facing signing agreements to vary their appointments to the University in this way are naturally concerned about what would happen to their employment conditions if a judgement is made at some future date that they should not continue in the Institute.[29]

The solution recommended was to 'second' staff for an initial period of five years research in the institute, with the person's research quality, productivity and contribution to its interdisciplinary programs to be assessed after three years by the institute's Scientific Advisory Committee with peer and international input.

It is recommended that all teaching and research staff conducting research in the Institute retain full-time appointments in their academic departments, seconded to the Institute for a specified period (initially five years) for the purpose of conduct of their research. Renewed secondment for rolling three-year terms would commence after the first three years, subject to review with peer and international components, of both the quality of their research and their contribution to the developing interdisciplinary programs of the Institute.

A small number of research-only staff may be appointed primarily within the Institute.

Where a faculty or department contributes a minor portion of such an appointment, this would be reflected in any departmental fraction of the appointment. If the majority of funding is from outside the faculty/department structure, the primary University appointment would be in a new Academic Organisational Unit within the Institute, headed by the Director … Visiting academic staff might also be granted short term appointments within such a unit.[30]

With the demolition of the Russell Grimwade Building, the Mab Grimwade Library, which had been named to honour Lady Grimwade following a council resolution in February 1958, was disbanded. The

The Russell Grimwade Building

library had been planned as far back as 1956, when a space previously designated as a staffroom had been reassigned for the purpose; book-shelves had been ordered in 1958. The Mab Grimwade Library was one of the responsibilities of Nancy Hosking and Evelyn Parkhill, among others, and before it was dispersed it accommodated some hundreds of books and journals, used by both staff and students. However, increased accessibility of scientific literature online made the accommodation of printed books and journals less important and less economically justifi-able. The only sign of the library's existence now is the original nameplate which is currently affixed to the wall in the foyer of the department in the Medical Building. On 30 June 2010, at a ceremony attended by Sir Andrew Grimwade and Fred Grimwade, the laboratories on the second floor of the South Tower of the Bio21 Institute were named the Grimwade Research Laboratories.

The move to the new premises, with much improved facilities for research, was of course welcomed by Biochemistry staff, but there were some regrets. These can be summed up by Bruce Livett's comment, which forms an interesting contrast to Trikojus's 1965 complaint about music classes:

> I loved the old building being on campus with its own library and proximity to the Conservatorium of Music, so I could hear the music students practising when I had the windows open in the summer and nip into a concert on a Monday lunchtime.[31]

The history of the Bio21 Molecular Science and Biotechnology Insti-tute in the David Penington Building on Flemington Road will be written by others, but it is appropriate briefly to describe its opening. Funds for the new development came from a multitude of sources, ranging from the Victorian and Australian governments to Atlantic Philanthropies and, of course, the university itself.[32] The institute was opened by the state premier, attended by the ministers for Innovation and Health, in the presence of the Vice-Chancellor and many leaders of Australian academia and industry. On that occasion, devoted to the celebration of a great step forward in collaborative research in biochemistry and molecular biology, no mention was made of Victor Trikojus's dream of a research and teaching department under a single roof as the speakers and audience recognised that research facilities needed to be shared among many disciplines and, most notably, needed to be financed by government and industry as well as academia.

Eight notable scientists attended 'Passionate Minds: A Colloquium of Women in Medical Research' in 2012. Standing, left to right: Pip Pattison, Ingrid Scheffer, Elizabeth Blackburn, Elizabeth Hartland, Leann Tilley
Seated, left to right: Jane Gunn, Judith Whitworth, Ruth Bishop

CHAPTER 7

Stars Make Other Stars

To devote most of a chapter to men and women who spent only a short time, generally at the beginning of their careers, in the department may seem to be a case of basking in reflected glory. To some extent this is true. It is, however, equally true that upon its establishment, the Melbourne University Department of Biochemistry became internationally renowned, a place where biochemists from other states and countries were keen to work. Victor Trikojus and his successors, however, were also responsible for identifying and fostering the talent of men and women who went on to occupy significant positions in prestigious institutions, often receiving international acclaim for their achievements. In fact, during a period of about three decades, an extraordinary number of people who were to forge great careers and lead internationally renowned organisations passed through the department.

One reason why the early career of a scientist is so important is that scientific teaching and research are, of necessity, organised differently from teaching in the humanities and social sciences. In history or English, a potential scholar is likely to be identified through the quality of the written work submitted in the early years of his or her undergraduate career. In science, the undergraduate course is largely a matter of absorbing and integrating a large body of material through lectures and large group classes, and it is not until the final or honours year of a course, when the student joins a laboratory team, that the potential star researcher can be identified. Identifying talent and fostering it is one thing: being in a position to offer a career path to students who display it is another, and senior members of the department from the time of Trikojus onwards had long been aware of Melbourne's limitations in this area. Trikojus had been unable to find a position for Peter Springell. Both Jean Jackson and Joseph Lugg had found better opportunities for advancement and research in Singapore. Pamela Todd, Francis Hird and Lloyd Finch all wrote with amazement and envy of the funds and facilities available for research in the United States. The rapid expansion of the Australian university sector from the 1960s onwards also provided many positions elsewhere in Australia to which Melbourne graduates were appointed.

One molecular biologist who took his first degrees from La Trobe and spent ten years in the Melbourne department through the mid-stage of his career, before moving to Monash, provides an interesting view of the value of work experience outside Australia. Professor Trevor Lithgow (1964–) was an Associate Professor in the Grimwade School of Biochemistry and Molecular Biology from 1999 to 2008, and before his appointment was the Australian federation fellow in the School of Biomedical Sciences at Monash. He received an Australian Laureate Fellowship from the Australian Research Council in 2013 and is an Honorary Professor in the Melbourne University Depart-

Trevor Lithgow, Professor of Biochemistry in his laboratory at Monash University

ment of Microbiology and Immunology. From 1993 to 1995 Lithgow was a research fellow at the Biozentrum of the University of Basel, working with Gottfried Schatz to determine the mechanisms that drive the protein import pathway in mitochondria. Writing of working overseas, he noted that:

> It forced me to 'grow up' as a scientist and it made me feel like a scientist – not a recently graduated student, not an apprentice in training, but part of the international enterprise that is science. This will sound elitist, but science is an elite sport. *Living the dream* in Europe also allowed me to drop any insecurities that I'd had about whether or not I was as good as the postdocs that student-me had imagined working at places like Princeton, Cambridge etc. We don't talk about this much, but I suspect that many Australian students have a similar cultural cringe, however unwarranted it is. You can tell yourself you're as good as the best, other people can tell you this, but if you want to be certain that your PhD really did bring you up to scratch – to an internationally competitive standard of excellence – there is nothing like working for a few years shoulder to shoulder with the best of your peers from around the world.[1]

The establishment of Monash University in 1961 attracted Joseph Bornstein to the foundation Chair of Biochemistry. That of the Research Schools of Chemistry and of Biological Sciences at the ANU in 1967 was to see Michael Birt take the position of inaugural Professor and Head of the Department of Biochemistry. La Trobe University enrolled its first students in 1967 and Bruce Stone took up his appointment to its foundation Chair of Biochemistry in 1972. As was to be expected, these men attracted postgraduate students from the University of Melbourne, and the well-established tradition of undertaking postgraduate work outside Australia meant that other stellar Melbourne graduates took their talents even further afield.

Lindsay Michael Birt (1932–2001) moved twice from Australia to the United Kingdom before settling into a career as an Australian university administrator. He took his BAgrSc in 1953, his BSc the following year and his PhD in 1957 from Melbourne, before winning the 1851 Exhibition to Oxford from which he took his DPhil. On his return to the University of Melbourne in 1960, he took a position as lecturer, was promoted to senior lecturer and left in 1964 to take a senior lectureship at the University of Sheffield, which position he occupied until 1967 when he was invited to the chair at the ANU. From 1973 his career, first at the University of Wollongong and subsequently at the University of New South Wales, was that of Vice-Chancellor, from which positions he oversaw considerable development and change in the university sector in general and his own institutions in particular. A second edition of *The Biochemistry of the Tissues* (first published in 1968) appeared in 1976, but although he continued to publish in biochemical journals, Michael Birt's main contribution to the published literature was principally in the field of educational administration and policy.[2] From his time in Melbourne, Beverley Bencina particularly remembers his lecturing skills.[3]

Birt oversaw the establishment of the University of Wollongong and its amalgamation with the Wollongong Teachers' College. At the University of New South Wales, he guided the establishment of the Australian Defence Force Academy, which had its genesis as that university's Faculty of Military Studies at the Royal Military College Duntroon.

Michael Birt (*left*) Vice-Chancellor of the University of New South Wales, with Ray Martin, Vice-Chancellor of Monash University, 1986

The Australian Defence Force Academy opened in 1986. Under its aegis, cadets and midshipmen undertake an undergraduate course concurrently with their military and leadership program. This means that at the conclusion of their training, as well as specialised skills, they have acquired an internationally recognised degree.

One person whose name has become famous in molecular biology left the department immediately after taking his doctorate under Francis Hird and Michael Birt. Lynn Dalgarno (1935–) took his BAgrSc in 1958 and his PhD on respiratory metabolism and processes of uptake in a plant tissue in 1962. Having won the A.M. White Scholarship, the Sir Arthur Sims Scholarship and a CSIRO Junior Research Scholarship in 1958, Dalgarno was awarded the Sir John and Lady Higgins Research Scholarship for 1958 to 1961. He was the editor of the Melbourne University Science Club's *Science Review* in 1959, when it resumed publication after a hiatus of seven years. In 1963 he travelled on a University of Melbourne Travelling Scholarship to London to join E.M. Martin in the Biochemistry Division of the National Institute for Medical Research. In 1965, with an MRC-CNRS Exchange Scholarship, he worked with François Gros at the Institut de Biologie Physico-Chimique in Paris, and in 1967, assisted by a US Public Health Research Grant, he worked in the Biology Division of the California Institute of Technology with Robert L. Sinsheimer.

On his return to Australia, Lynn Dalgarno took up an appointment at the ANU, first as senior lecturer and subsequently as reader. His research was in two main areas. The first was a study of mosquito-borne RNA viruses, including Ross River virus (RRV) and Murray Valley encephalitis virus (MVE). Dalgarno and his collaborators explored virus replication in vertebrate cell culture and the establishment of persistent virus infection in mosquito cells. Using recombinant DNA techniques adapted to single-stranded viral RNA, they demonstrated that these viruses show little genetic difference between isolates from widely separated geographic regions, although stable genetic subtypes did exist. RRV genetic stability was high during epidemic spread among non-immune human populations, implying very limited virus evolution under these conditions. Dalgarno and his colleagues determined the complete nucleotide sequence of the RRV and MVE RNA chromosomes and partial nucleotide sequences for yellow fever, Barmah Forest and dengue viruses.

Lynn Dalgarno at Caltech in 1981

To understand the relationship between the biological/pathogenic properties of the virus and the precise nucleotide sequence of the RNA, they
used natural variants and laboratory-derived mutants of RRV and MVE
to identify viral virulence determinants in mice. They studied the replication of RRV and yellow fever virus genetic variants to identify antigenic
sites and neutralisation determinants in the viral coat proteins.

His second area of interest was ribosomal RNA (rRNA). Initially
Dalgarno studied the biosynthesis and physico-chemical properties of
insect rRNA. In most insects the large (26S) rRNA splits into two roughly
equal fragments on heating, unlike other rRNAs. He and his PhD student,
John Shine, used 3'-terminal labelling and nucleotide sequence analysis of
the rRNA to explore this. Surprisingly, they showed that the 3'-sequences
of the small 18S rRNA from widely divergent eukaryotes (*Drosophila*,
yeast and rabbit reticulocytes) are identical (GAUCAUUA-OH) and thus
highly conserved during evolution, implying an important function.
Dalgarno then realised that this sequence contained the complement of
the three termination codons (UAA, UAG and UGA) and could therefore potentially base pair with them. From this, he and Shine proposed a

model for the termination of protein synthesis in which RNA:RNA base pairing between ribosomal RNA and termination triplets took place.

The idea that rRNA:mRNA base pairing might be involved in protein synthesis led to sequence studies on the small 16S rRNA of a number of bacterial species. From these, Dalgarno and Shine proposed that the 3'-terminus of 16S rRNA interacted with prokaryotic mRNA to help identify the correct initiator AUG among the many internal AUGs in the mRNA. A series of papers followed on the role of the 3'-end of 16S rRNA in the initiation of protein synthesis in bacteria, the most frequently cited of which was published in 1974.[4] It proposed a mechanism by which bacterial ribosomes select the correct site for the initiation of protein synthesis on messenger RNA. Dalgarno and Shine proposed that in *Escherichia coli* the ribosome recognises the correct initiation triplet AUG on mRNA as the result of base pairing between a short pyrimidine-rich stretch of nucleotides (PydACCUCCUUA-OH) at the 3'-end of the small 16S rRNA, and a complementary purine-rich sequence (AGGAGGU) in the mRNA around eight nucleotides upstream of the initiator codon. This led to the proposal that the precise nucleotide sequence at the 3'-end of the rRNA determines the intrinsic capacity of the prokaryotic ribosome to translate a particular cistron in mRNA. The degree of complementarity between the 3'-end of the rRNA and the sequence preceding the initiator AUG provided a mechanism by which the cell distinguished the correct initiator AUG from numerous internal and out-of-phase AUGs in the mRNA. These insights were also important in optimising the large-scale expression of synthetic mammalian genes in bacterial cells.

In 2010 John Shine (1946–), in whose honour the Shine Dome in Canberra was renamed, was awarded the Prime Minister's Prize for Science. The Shine–Dalgarno sequence was a key discovery allowing further development of molecular biology. From 1997 to 2003 Dalgarno worked part-time as a consultant in intellectual property cases involving molecular biology and molecular virology, and since 2003 he has found the time to return to playing the viola in various chamber music groups and orchestras.

Lawrence Walter Nichol (1935–) is another Vice-Chancellor who began his postgraduate career in the department. After taking his BSc (Hons) in 1957 and his PhD in 1962 from the University of Adelaide –

which also awarded him a DSc in 1974 – he worked there from 1959 to 1960 on a postgraduate fellowship. He spent 1961 and 1962 at Clark University, Worcester, Massachusetts as a postdoctoral Fulbright fellow, and was a research fellow at the ANU from 1963 to 1965. Nichol joined the Melbourne Biochemistry Department in 1966 as a senior lecturer, the same year he won the David Syme Research Prize. He was to work in the department (he was promoted to reader in 1969) until he returned in 1971 to the ANU as Professor of Physical Biochemistry, in which position he remained until 1985 when he was appointed Vice-Chancellor of the University of New England. From 1988 to 1993, Nichol was Vice-Chancellor of the ANU.

Among his most-cited papers published during his time at Melbourne is a theoretical study of the binding of small molecules to a polymerising protein system. Another paper on the molecular weight and stability of concanavalin A, based on work in the department, was published shortly after he took up his appointment at the ANU.[5]

Lawrence Walter Nichol, Vice-Chancellor from 1988 to 1993, and Emeritus Professor at the Australian National University

Bruce Stone photographed in 1981 when he was Foundation Professor of Biochemistry at La Trobe University

Bruce Arthur Stone (1928–2008) spent over a decade in the Melbourne department before taking the position of Professor of Biochemistry at La Trobe, which he occupied for more than twenty years. A graduate of the University of Melbourne (BSc, 1948), Stone worked for two years on ways of preserving canvas from tropical moulds at the Department of Supply laboratories in Maribyrnong before going to Kew Gardens in London. There, he enrolled for a PhD on the enzymes involved in polysaccharide metabolism, which he was awarded in 1954. Returning to Melbourne after postdoctoral work in Ottawa and London, Stone was appointed to a lectureship in 1958 and a senior lectureship in 1963, before being promoted to reader in 1966. In 1972 he left Melbourne to become the first Professor of Biochemistry at La Trobe. During his long academic life, Stone published papers on the chemistry and biochemistry of carbo- hydrates, especially the polysaccharides, with a succession of Melbourne and La Trobe biochemistry students who became luminaries in their own right: among the earliest was his first PhD student, Adrienne Clarke.[6] Other recruits to the department at La Trobe, such as Dick Wettenhall and

Leann Tilley, would also go on to distinguished careers in Biochemistry at Melbourne. At the time of his death, Stone was working with Adrienne Clarke on updating their 1992 study on glucans.[7]

Bruce Stone was an active member and strong supporter of many professional organisations, which in turn recognised his work. He joined the Australian Biochemical Society (forerunner of the Australian Society for Biochemistry and Molecular Biology) in the early 1960s, becoming its president from 1988 to 1990. In 1985 he won the F.B. Guthrie Award for Cereal Chemistry of the Royal Australian Chemical Institute, and in 2004 the Thomas Burr Osborne Award of the American Association of Cereal Chemists. In 2000 he was made a corresponding member of the American Association of Plant Physiologists and in 2001 he became an ISI Australian Citation Laureate.[8] His most frequently cited paper, co-written with three La Trobe colleagues, described a simple and rapid method for the preparation of alditol acetates from monosaccharides that could be performed in a single tube without transfers or evaporations. In it, monosaccharides are reduced with sodium borohydride in dimethyl sulphoxide and the resulting alditols acetylated using 1-methylimidazole as the catalyst. Removal of borate is unnecessary and acetylation is complete in ten minutes at room temperature. Monosaccharides are quantitatively reduced and acetylated by this procedure. The alditol acetates are completely separated by glass-capillary, gas-liquid chromatography on Silar 10C. The investigators had applied the method to the analysis of monosaccharides in acid hydrolysates of a plant cell-wall.[9]

Adrienne Clarke (1938–) AO, AC, FAA, FTSE is, as the *Encyclopedia of Australian Science* spells out, a scientist of firsts. Her team was the first to clone the gene (S gene) which regulates self-compatibility in plants and the first to clone the cDNA of

Adrienne Clarke, botanist, biochemist and Chancellor of La Trobe University an arabinogalactan protein, which is a class of plant proteoglycan. She was the University of Melbourne's first female laureate professor and the first woman appointed to the office of lieutenant governor of Victoria, a position she occupied from 1997 to 2000, and which from 2006 has been occupied by another woman, Marilyn Warren.[10] Chair of the Board of CSIRO from 1991 to 1996, Clarke is a fellow of both the Australian Academy of Science and the Australian Academy of Technological Sciences and Engineering. She is a foreign associate of the National Academy of Sciences (USA), a foreign member of the American Academy of Arts and Sciences, and a fellow of Janet Clarke Hall in Melbourne.

Adrienne Clarke took her BSc (Hons) in 1958 and her PhD in Biochemistry in 1962. She is listed among the full-time research workers in Biochemistry from 1958 to 1961, but most of her work after five years overseas was in the Botany Department, where she has occupied a variety of positions since 1974. Clarke continued, however, to publish with colleagues in the discipline of biochemistry at the university, including, in 1979, a future head of department, Paul Gleeson. He was at the time a graduate student funded by CSIRO.[11]

Another stellar biochemistry graduate is also a scientist of firsts: the first female director of the WEHI from 1996 to 2009 was also the first Australian winner of the L'Oréal Unesco Award for women in science and the first Australian woman to win the royal medal for science from the Royal Society. Suzanne Cory (1942–) AC, FAA, FRS took her BSc in 1964 and her MSc the following year from the University of Melbourne, and began working with Fred Sanger towards the PhD which she was awarded in 1968 at the MRC Laboratory of Molecular Biology at Cambridge University. Her PhD work determining the sequence of transfer RNA led to postdoctoral sequence

Suzanne Cory, the first woman president of the Australian Academy of Science, in her laboratory

analysis of R17 bacteriophage RNA as a model messenger RNA at the University of Geneva. In 1971, Cory and her husband, Jerry Adams, began an enduring and productive partnership at the WEHI. Suzanne Cory's awards and honours range from the Lemberg Medal of the Australian Society of Biochemistry and Molecular Biology to the Companion of the Order of Australia. Suzanne Cory High School, one of four selective-entry state high schools in Victoria, opened in 2011.

Confirming that the Melbourne department provided a good training ground, Nicholas Johannes (Nick) Hoogenraad (1942–) commented in an interview in 2010 that 'Frank Hird's lab was where just about all the professors in biochemistry in Australia came through in the early days'.[12] Hoogenraad is one of them. He took his BSc in 1965 and his PhD in 1969. His thesis on the contribution of rumen bacteria to the nutritional requirements of sheep was influential. During his PhD work with Hird, Hoogenraad came into contact with Ian Hamilton Holmes (1935–2010), the notable electron microscopist, with whose help he and Hird made the first atlas of bacteria from the rumen. Hoogenraad noted in 2010 that, five years after submitting his thesis, he was still being asked to talk about the work he did as a PhD student. In an interesting echo of Ernest Rutherford's oft-quoted remark that 'we had no money so we had to think', Hoogenraad recollected:

I had come out of Frank Hird's lab, where I had had a very strict education. I also had a very Australian education, in the sense that we still had a workshop just for postgraduate students in biochemistry at Melbourne University and I had learnt to make my own equipment. Frank privately tutored all of his students in how to do glass-blowing and make equipment. Early on in my life, like many Australians, I learnt how to fix my own car. For example, I once had to change the clutch in my Volkswagen and things like that. When you went to America in those days you were quite unusual for being handy with your hands.[13]

Demonstrating the close relationship between the departments of Biochemistry and Microbiology, Hoogenraad's first published paper was co-authored with Ian Holmes, Francis Hird and Nancy Millis.[14] After some time overseas, Hoogenraad returned to Australia to the

Department of Biochemistry at La Trobe University, where his research has focused on mitochondrial biogenesis and on the role of molecular chaperones in protein targeting and folding in mammalian cells. He is also notable for his work promoting science in schools, and he was awarded the Pharmacia-LKB Medal of the Australian Society of Biochemistry and Molecular Biology in 1994, the Leach Medal in 1997 and the Lemberg Medal in 2004. He is the executive director of the La Trobe Institute for Molecular Science, which opened on 15 February 2013: as well as housing thirty-four research and

Nick Hoogenraad opening the International Biochemistry and Molecular Biology Conference in 2010

eleven teaching laboratories, it has over 400 research, teaching and other staff. The Hoogenraad Auditorium is named in his honour.

The decades from the late 1960s to the late 1980s were extraordinary for the number of Biochemistry graduates who went on to head research centres all over Australia and overseas. In a restricted space, it is possible to summarise the careers of only a few. Many of these scientists maintain strong personal and professional links with the Melbourne department.

One of Australia's pre-eminent cancer researchers, Antony (Tony) Wilks Burgess (1946–), took his BSc (Hons) in 1969 and his PhD, under the supervision of Syd Leach, on 'Chemistry of Glyceraldehyde-3-phosphate Dehydrogenase from *Jasus verrauxi*; Conformational Energy Calculations on Methylated Peptides' in 1974. After postdoctoral work in the Chemistry Department of Cornell University and the Biophysics Department of the Weizmann Institute in Rehevot, Israel, with Professor Harold Scheraga, he took a position in 1975 as a postdoctoral research fellow at the WEHI, where, from 1977 to 1979, he was senior research officer and head of the Laboratory of Biological Regulators. He became director of the Ludwig Institute for Cancer Research from 1980 to 2009, combining this with various senior positions in the departments of Medicine and Surgery at the University of Melbourne. Since 2012 he has headed a laboratory in the Structural Biology Division of the WEHI that studies the biology and biochemistry of normal and cancerous intestinal glands (crypts).

Antony Burgess, laboratory head in the Structural Biology Division of WEHI

Tony Burgess, who was made a Companion of the Order of Australia in 1998 and received a Centenary Medal in 2001, has won numerous awards, ranging from the Gottschalk Medal from the Australian Academy of Science in 1981 to the President's Award of the Cancer Council Victoria in 2010 and the Clunies Ross Award of the Academy of Technological Sciences and Engineering in 2011. He has published over 250 research papers. Among his major discoveries are the identification of G-CSF, the purification of GM-CSF, the analysis of the 3D structure and function of cancer-associated EGF receptors, and the role of wnt signalling in colon stem cell–cell adhesion.[15]

Geoffrey Bruce Fincher (1946–) took his BAgrSc from the University of Melbourne in 1967 and his PhD – 'Studies on Polysaccharides in Wheat Endosperm' – in 1973, leaving, as so many postgraduates do, for post-doctoral work overseas. After two years in England as research fellow at the Brewing Research Foundation in Surrey, he spent the following two as a teaching postdoctoral fellow at McGill University, Montreal. Returning to Australia in 1977, he joined Bruce Stone's Department of Biochemistry at La Trobe University, becoming, in 1988, reader in plant biochemistry. In 1993, he took up the position of Professor of Plant Science at the University of Adelaide, Waite Campus, becoming director of the Waite Agricultural Research Institute in 2003. From 2003 to 2010 he was deputy CEO of the Australian Centre for Plant Functional Genomics, which is based at the University of Adelaide. He is currently the director of the $32 million Australian Research Council Centre of Excellence in Plant Cell Walls, which received funding to take it from 2011 to 2017.

Geoffrey Fincher won the 2000 F.B. Guthrie Award, which is the highest award of the Royal Australian Chemistry Institute's Division of Cereal Chemistry and honours the contribution of the pioneer cereal chemist, F.B Guthrie, to wheat research in Australia. In 2010 he won the American Society of Plant Biology Corresponding Membership Award. This honour, initially given in 1932, provides life membership and society publications to distinguished plant biologists from outside the United States. He has been a member of the editorial boards of numerous professional journals, including *Planta*, *Plant Physiology* and *Seed Science Research*.

Geoffrey Fincher, Australian Research Council Centre of Excellence in Plant Cell Walls

Publications from Fincher's laboratory have been cited over 10 000 times. Two of his most-cited recent publications are 'Cellulose Synthase-like *CslF* Genes Mediate the Synthesis of Cell Wall (1,3;1,4)-β-Glucans' and 'An *Arabidopsis thaliana* Callose Synthase, GSL5, is Required for Wound and Papillary Callose Formation'.[16] He was a member of the International Barley Genome Sequencing Consortium, which published a physical, genetic and functional sequence assembly of the barley genome in *Nature* in 2012. He is a member of the board of the A.W. Howard Memorial Trust of SARDI, the South Australian Research and Development Institute. In 2013 Fincher received the Thomas Burr Osborne Medal, which is named after the great protein chemist and is the most significant medal awarded by the American Association of Cereal Chemists International. The award was established in 1926 and recognises distinguished contributions in the field of cereal chemistry. At the award ceremony, Geoffrey Fincher also learned that he had been accepted as a fellow of the association.

Matthew Anthony Perugini (1972–) moved from Melbourne to La Trobe, spent his undergraduate years and the first decade and a half of his professional career in the Department of Biochemistry and

Molecular Biology, and has held an honorary position there since 2012. Perugini, whose research is directed to the structure, function, regulation and inhibition of enzymes of the lysine biosynthesis pathway in bacteria, plants and lower eukaryotes, and the development of frontier technologies for measuring biomolecular interactions, such as protein–protein and protein–drug interactions, took his BSc (Hons) in 1994 and his PhD in 2001 in Geoff Howlett's laboratory.[17] After graduation in 2002, having worked as a demonstrator, tutor and senior tutor since 1994, Perugini was appointed to a continuing position as senior tutor and assistant coordinator of the relatively new Bachelor of Biomedical Science course which had been established in 1999: from 2003 to 2009 he was lecturer and then senior lecturer in this course. The BBiomedSc course was ranked in the top seven science courses by an international panel of academics in a 2006 survey. Promoted in 2009 to associate professor and reader, Perugini took up a position as Associate Professor in the La Trobe Institute for Molecular Science, within that university's Department of Biochemistry.

Perugini has won almost as many awards as a teacher as for his research. He was a two-time winner of the Outstanding Lecturer Award, as voted by the BBiomedSc students in 2004 and 2009, the only two

Matthew Perugini and the Fast Protein Liquid Chromatography machine in the Howlett Laboratory

occasions on which this award has been bestowed. His research has been recognised through many awards, including the Young Biophysicist Award of the Australian Society for Biophysics in 2004, the Australian Society for Biochemistry and Molecular Biology's ABI Edman Award in 2007, the Erskine Fellowship of the University of Canterbury in New Zealand in 2011 and the Excellence in Research Award from La Trobe University in 2012. His publications as laboratory head in the Biochemistry Department include a seminal paper reporting the structure and function of the first non-tetrameric DHDPS enzyme to be discovered that offers an insight into the molecular evolution of enzyme quaternary structure. The paper also reports one of the first applications of the fluorescence detection system developed for the analytical ultracentrifuge, which offers exquisite sensitivity (down to picomolar concentrations of analyte/sample).[18] Perugini lists travel, reading and cooking among his hobbies as well as several sports, including surfing. In 2013 he was an inaugural co-leader of the Chemical and Biological Defence theme of the recently established Defence Science Institute in Melbourne, and is treasurer of the Lorne Conference on Protein Structure and Function. He has been a member of the Richmond Football Club since 1996, with Gold Member status since 2005.

The year after Nick Hoogenraad took up his position at La Trobe saw the departure for Cambridge University of a young female graduate who was set for a truly stellar career, including sharing the 2009 Nobel Prize in Physiology or Medicine with Carol W. Greider and Jack W. Szostak for the discovery of how chromosomes are protected by telomeres and the enzyme telomerase. Elizabeth Helen Blackburn (1948–), the first Australian female Nobel laureate, was awarded her BSc (Hons), supervised by Theo Dopheide and Barrie Davidson, in 1970 and her MSc under the supervision of Francis Hird in 1972. Her BSc work led to her first published paper, co-authored with Davidson and Dopheide.[19] Her MSc on the metabolism of glutamine in the liver examined how glutamine was changed into glutamate, thereby releasing ammonia. This also provided material for a number of papers.[20]

Hird introduced her to the double Nobel laureate Frederick Sanger (1918–2013) and recommended her as a research student, later describing her as having 'about the best integrating mind I have ever seen'.[21]

Blackburn began work in Sanger's Laboratory of Molecular Biology at Cambridge late in 1971. After postdoctoral research at Yale, she worked at the University of California, Berkeley and subsequently the University of California, San Francisco. Her career after her time at Melbourne has included fellowship of the Royal Society of London, the American Academy of Arts and Sciences, and the Australian Academy of Sciences, as well as numerous awards for cancer research, genetics and the biomedical sciences generally, and several honorary doctorates. Blackburn discovered the ribonucleprotein enzyme, telomerase, and investigates the effect that the manipulation of telomerase activity has on cells. Her laboratory work is directed towards elucidating the biology of telomerase and telomere. Quoting Blackburn's Nobel citation, Jane Carey comments that:

> This muted description does little justice to the significance of these discoveries. The telomere is a cap-like structure at the end of chromosomes that stops them from disintegrating during cell division. Telomerase is the enzyme that builds them. We literally could not live without them. Blackburn's research uncovered a fundamental biological process and she founded an entirely new scientific field.[22]

Elizabeth Blackburn is a past president of the American Association for Cancer Research and the American Society for Cell Biology, a Companion of the Order of Australia, and was a member of the US president's Advisory Council on Bioethics from 2001 until her dismissal in 2007, after her support for human embryonic cell research brought her into conflict with the Bush White House.

During a visit to the University of Melbourne in 2012, Blackburn spoke of the special place of women in scientific research: she believes that they may bring a distinctive approach to

Elizabeth Blackburn, Nobel laureate, photographed 12 April 2012 at Chemical Heritage Day at the Chemical Heritage Foundation, Philadelphia, PA, USA

problems. In her own research she has investigated the effects of severe, chronic and drawn-out stress on telomeres, finding quantifiable correlations between the severity of stress experienced and the shortness of telomeres. These findings should provide a powerful incentive to funding bodies to direct more resources towards behavioural and social sciences research.[23] The Elizabeth Blackburn School of Sciences, located within the Bio21 Institute, was launched in 2014. It provides access for 200 Victorian Year 11 and 12 students in science, technology, engineering and maths, and aims to foster science education at both the school and university level in a way that connects 'school to bench to workplace'.[24]

Mary-Jane Hilliard Gething (1949–) does not really belong in the category of stars associated for only a short time with the department, although she spent much of her career overseas. Gething was a contemporary of Elizabeth Blackburn and they shared a house in Drummond Street, Carlton as well as their laboratory as undergraduates. Despite more than twenty years in Britain and the United States, however, Gething spent far more time in Biochemistry at Melbourne, including appointments as reader and Associate Professor in Biochemistry from 1994 to 1998, and as Professor of Biochemistry and Molecular Biology from 1998 to 2005. Like Blackburn, Gething took her BSc (Hons) in 1970: unlike her, she took her PhD also from Melbourne, graduating in 1973 with a thesis on structural and physiochemical studies on chorismate mutase-prephenate dehydratase from *E. coli*. In 1972 she became the first woman to win the Shell Science-Engineering Scholarship, in the first

Mary-Jane Gething outside the Carlton house she shared with Elizabeth Blackburn

year in which it was opened to women, and used it for postdoctoral study at the Medical Research Council Laboratory of Molecular Biology at Cambridge. She made her mark at the University of Melbourne in other ways as well, with her election in 1971 as the first female postgraduate student representative on the University Council.[25] She was also a founding member of the Women's Electoral Lobby and Community Control Childcare. Professor Mary-Jane Gething was also the first woman to head the department, chairing it from 2000 to 2005.

Having undertaken postdoctoral work at Cambridge, Gething worked on protein sequencing in London from 1976 until 1980, when she moved for what was to be the next decade and a half to the United States. She began working at Cold Spring Harbor Laboratory in 1982, where she continued her research of proteins. In 1985, Gething and her husband, Joseph Frank Sambrook (1939–), moved to Dallas to work at the University of Texas Southwestern Medical Center. She noted in an interview in 2003 that when she came back to Melbourne there were fewer women in her department than when she was a student, and commented that her return to Australia was largely made possible by the fact that Sambrook had taken his PhD in Australia and was happy to relocate, taking up the position of director of research at the Peter MacCallum Institute.[26]

Their scholarly partnership proved extremely productive, with over a dozen jointly authored papers appearing during the 1980s and 1990s, among the most cited being on protein folding in the cell. They found that in the cell, as in vitro, the final conformation of a protein is determined by its amino-acid sequence, but whereas some isolated proteins can be denatured and refolded in vitro in the absence of other macromolecular cellular components, folding and assembly of polypeptides in vivo involves other proteins, many of which belong to families that have been highly conserved during evolution.[27]

The third woman in a quartet destined for outstanding lives in biochemistry who took their PhDs in the early 1970s has devoted much of her career to making a significant impact on the health of all Australians, but most notably the Indigenous community. While Jean Jackson did her early research work, and took her PhD, in Singapore, Audrey Cahn, who succeeded her in the department, did all of her research in Victoria. Kerin O'Dea (1946–), who took her BSc (Hons) in 1967 and her PhD, entitled

'Physico-chemical Studies on the Lactate Dehydrogenase of *Streptococcus cremoris*', in 1971, did much of her most important work geographically between the two of them. After time in the Research Division of the Cleveland Clinic in Ohio, Bayer in Germany and INSERM in Paris, O'Dea found her research interest turning from drug design to diabetes and obesity. She spent part of the period between 1976 and 1988 in various research positions at the Royal Melbourne Hospital, Baker IDI and Alfred Hospital and the Repatriation General Hospital (University of Melbourne). From 1988 to 1998 she worked at Deakin University as Professor of Human Nutrition, then as Dean of the Faculty of Health and Behavioural Sciences from 1993 to 1995, and as Pro Vice-Chancellor (Research) from 1995 to 1996. Two years at Monash as Professor of Nutrition and Preventive Medicine and head of the Centre for Population Health and Nutrition followed, and then, as reported in *Australian Doctor*, the 'O'Dea hurricane' headed north.[28]

Kerin O'Dea was appointed to head Australia's only medical research institute dedicated to improving Indigenous health and wellbeing: the Menzies School of Health Research. She had done significant work in the field of Indigenous health for over twenty years. She is probably best known for her novel research on the marked beneficial health impact of temporary reversion to a traditional hunter-gatherer lifestyle on diabetes and associated conditions in Indigenous Australians.[29] In this study, conducted in 1982, she accom-

Kerin O'Dea has worked in Indigenous health for two decades

panied a group of diabetic Indigenous people back to their traditional country in the Kimberley, where they lived for seven weeks as hunter-gatherers. By the end of this period, 'the major metabolic abnormalities of type II diabetes were either greatly improved or completely normalized. At least three factors known to improve insulin sensitivity (weight loss, low-energy diet, and increased physical activity) were operating in this study and would have contributed to the metabolic changes observed'.[30] This study was life-changing, and drove O'Dea's subsequent research focus on the therapeutic and preventive potential of traditional diets and lifestyles, such as the hunter-gatherers of Australia and the Cretans in Greece in the 1950s.

After five years in Darwin – during which her work was recognised in 2001 with a Centenary Medal for service to research in Australian Indigenous health, and in 2004 with an Order of Australia for service in the areas of medical and nutrition research to the development of public health policy, and to the community, particularly Indigenous Australians, through research into chronic disease and prevention methods – O'Dea became director of the Sansom Institute for Health Research at the University of South Australia from 2009 to 2012: she took up the position of Professor: Nutrition and Public Health in the university's Health Sciences Division in 2013. She is a board member of Outback Stores, which was established in 2006 to provide retail services to remote stores on behalf of Indigenous communities.[31] Among her most recent publications is a paper on the effect of the Mediterranean diet on the health of people with non-alcoholic fatty liver disease.[32]

The last woman in a constellation of Melbourne biochemistry PhDs who pursued their careers outside the department, took her doctorate in Biochemistry and Microbiology, worked for many years in the Botany School and then, after a period of research in industry, returned to a career in university administration at Monash. Edwina Cecily Cornish (1955–) took her PhD, on the TyrR gene of *Escherichia coli* K-12, in 1984. While remaining at the University of Melbourne until 1988, she worked in the School of Botany, moving subsequently to Florigene Ltd, an Australian biotechnology company that uses gene technology to develop new flower varieties for

Edwina Cornish, inaugural Provost and Senior Vice-President of Monash University

the cut-flower market. Cornish began her research at the moment when the study of molecular biology transformed biochemical studies, and her work was significant in the genetic modification of plants. The first aim was to create a blue rose, and in 1991 the Florigene team successfully isolated the 'blue gene' from the petunia flower, filing patents in all major countries in 1992. In 1994 the 'blue gene' was implanted into carnations, and mauve and later purple varieties came onto the market in 1996 and 1997. The process involved is described by Florigene:

> Anthocyanins are the most prevalent class of pigment compounds in plants; in flowers these are deposited in the large vacuoles of petal epidermal cells where most flower colour is localised. Anthocyanins are relatively simple, water soluble, structures with the hydroxylation pattern of the B ring at positions 3', 4' and 5' being a key colour determinant. Further modifications to the molecule are common and beyond imparting small changes to max lend stability to molecular complexes involving anthocyanin molecules, copigment molecules (typically flavonols, a product of the same pathway) and in notable cases metal ions. Such complexes are further influenced by petal epidermal vacuolar pH. All of these components impact directly on the observed colour.
>
> A key branch point of the anthocyanin biosynthetic pathway is centred around the intermediate DHK (dihydrokaempferol). The enzymes FLS (flavonol synthase), F3'H (flavonoid 3' hydroxylase), F3'5'H (flavonoid 3'5' hydroxylase) and DFR (dihydroflavonol reductase) all utilise this substrate, suggesting that this region of the pathway is a key determinant of anthocyanin biosynthesis and thus flower colour.
>
> Florigene/Suntory have cloned and protected nearly all genes in the anthocyanin/flavonoid pathway from petunia and numerous homologues from a diverse number of species.[33]

In 2000 Cornish moved to the University of Adelaide and, while continuing her research as Professor of Biotechnology, took on the position of Deputy Vice-Chancellor (Research). At the beginning of 2004 she was appointed Deputy Vice-Chancellor (Research) at Monash University and in August 2009 she was also appointed Senior Deputy Vice-Chancellor. In September 2012 Edwina Cornish became the inaugural Provost and Senior Vice-President of Monash University. This position is the Vice-Chancellor's core academic deputy, responsible for the management, supervision and control of faculty operations and academic portfolios related to faculty.

One Biochemistry postgraduate who pursued a career in dental science was Eric Charles Reynolds AO. He took his BSc (Hons) in 1973 and then his PhD, with a thesis entitled 'Thymidine Sensitivity of Murine Myeloma and Lymphoma Cells', and embarked on research into dental caries prevention and remineralisation, identifying molecular processes enabling the repair of early tooth decay. He subsequently led a team developing a range of Recaldent™ products, including a sugar-free chewing gum that substantially reduces the risk of tooth decay.

Eric Reynolds has received many awards for his work over the years, including the 1992 William J. Gies Award, presented by the International Association for Dental Research (IADR) and the American Association for Dental Research for the best paper published in the *Journal of Dental Research*; the Alan Docking Science Award, presented by the Australian and New Zealand Division of the IADR for outstanding scientific achievement in the field of dental research in 1997; and the Loftus Hill Memorial Medal Award of Merit of the Dairy Industry Association of Australia for outstanding scientific achievement in 2001.

The Victoria Prize is awarded by the Victorian Government to acknowledge excellence and recognise the contribution of scientists, engineers and innovators to the state's future. In an article reporting the awarding of a Victoria Prize to Eric Reynolds in 2005, *UniNews* noted:

> Professor Reynolds is one of the leading researchers in the University's Bio21 Molecular Science and Biotechnology Institute. He also heads the School of Dental Science and leads the CRC for Oral Health Science and the Victorian Centre for Oral Health Science.
>
> Congratulating Professor Reynolds on winning the Victoria Prize, Deputy Vice-Chancellor (Research), Professor Frank Larkins, said that in the quarter of a century Professor Reynolds has been involved in dental science he has made a number of key discoveries.

'He first identified the molecular processes enabling the repair of early tooth decay without the need for invasive treatment. This was followed by the discovery of a milk compound, called Recaldent™, that repairs the effect of acid on teeth and reduces the risk of disease.

'Recaldent is now used in gels, chewing gum, pastes and rinses around the world', Professor Larkins said.

Gippsland farmers provide the milk used to produce the protective compound.

Professor Reynolds describes Recaldent as a fundamental but important breakthrough. 'Dental decay costs Australians around $2 billion a year', he says.[34]

As well as publications in over 170 peer-reviewed journals and nineteen patents, Reynolds, who was appointed Head of the School of Dental Science in 1999 (having held a chair since 1994), was named as a laureate professor in 2011. Appointment as a Melbourne laureate professor represents a significant honour and the university undertakes to ensure that prominence is given to the recipients' work. They are appointed from among Nobel laureates, scholars equivalent in standing to Nobel laureates from fields in which Nobel Prizes are not awarded, or distinguished members of the existing University of Melbourne professoriate. In 2002 Reynolds won the Clunies Ross National Science and Technology Award, which recognises and honours people who have made noteworthy contributions to science and its application for the economic, social or environmental benefit of Australia.

Apart from Nick Hoogenraad, who ran away to sea for three months before returning to school at the age of fifteen, Robert Cornelis (Bob) Augusteyn (1941–) came to research by a more circuitous route than most of the other people who joined the department. Born in Amsterdam, he arrived in Australia at the age of nine with his parents in the great postwar influx of Dutch migrants. Having left school at fourteen, he completed his secondary education part-time, deciding only relatively late that he wanted to study chemistry, and taking a junior technical assistant position part-time in the Department of Biochemistry at the University of Queensland. He took his BSc (Hons) in 1965 and his PhD with a thesis entitled 'Some Studies in Protein Chemistry' in 1969. After a two-year

Robert Augusteyn,
director of the
National Vision
Research Institute

postdoctoral fellowship investigating α-crystallin, the principal protein structure of the lens, with Abraham Spector (1926–) at the Harkness Eye Institute at Columbia University, Augusteyn came to Biochemistry at Melbourne University initially as a research fellow and then, in 1972, as a lecturer. He was promoted to senior lecturer in 1980 and, following a year as Jess Cox Research Fellow at MIT, he was promoted to reader in 1988. From 1991 to 2002 he was director of the National Vision Research Institute.

Augusteyn's work on human cataracts and their causes won him the Shorney Prize for the most substantial contribution to knowledge in ophthalmology by an Australian over a three-year period, published in medical or scientific literature. The prize, awarded by the University of Adelaide, commemorates Herbert Frank Shorney (1878–1933), a graduate of the University of Melbourne who taught and practised in Adelaide.[35] Augusteyn's work on the growth and ageing of the lens and the structure of α-crystallin led to his being acknowledged as one of the leading experts in a field in which he published some three dozen papers; in 1988 he proposed what was to be accepted as the model for its structure.[36] Augusteyn's research was extremely successful in attracting funding from within Australia and overseas, with significant grants between 1979 and 1990 from the National Institutes of Health in the

United States, the Australian NHMRC, the Victorian Government and the Australian Retinitis Pigmentosa Foundation Pty Ltd.

Bob Augusteyn's work after leaving the Biochemistry Department has been described by Barry Cole in a paper to mark his retirement as director of the National Vision Research Institute.[37] The Vision Cooperative Research Centre, an international collaboration of centre researchers and industry bodies, was established in 2003 by the Australian Government. Its funding was extended for five years in 2010. While chairing its Scientific Committee, Augusteyn continues to publish on the human lens.[38] Bob Augusteyn is one of the authors of *The Eye*, a two-volume work published in 1979, and has written over 150 published papers. He is also a qualified international table tennis umpire.[39]

Richard Ian Christopherson (1949–) spent about eleven years in and out of the department before moving to the Department of Biochemistry at the University of Sydney in 1986, where he rose from lecturer to a Personal Chair in 1998 and was the foundation head of the newly established School of Molecular and Microbial Biosciences from 1998 to 2003. Christopherson took his BSc in 1970 and then his PhD in 1976 with 'Interrelationships of Pyrimidine Biosynthesis in *Escherichia coli*

Richard Christopherson researches cancer proteomics at the University of Sydney

K12, working with Lloyd Finch. He worked as a tutor in the Russell Grimwade School of Biochemistry from 1971 to 1972. After completing his PhD he was awarded a fellowship from the Damon Runyon–Walter Winchell Cancer Research Foundation for 1976–78. The foundation was established as the Damon Runyon Cancer Memorial Fund in 1946 by the broadcaster Walter Winchell, following the death of the writer Damon Runyon, and includes eleven Nobel Prize winners among its alumni. From 1978 to 1980 Christopherson was a special fellow of the Leukemia (now Leukemia and Lymphoma) Society of America. He returned to Australia as a research fellow in biochemistry in the John Curtin School of Medical Research at the Australian National University.

In 1983 Christopherson returned to the School of Biochemistry at the University of Melbourne as the C.R. Roper fellow in medical research. He investigated enzyme inhibitors which he had designed to kill cancer cells selectively by blocking early steps in the pyrimidine pathway; he also analysed the effects of three enzyme inhibitors under investigation at the US National Cancer Institute. His team developed the concept of concurrent use of two or more inhibitors that bind to different enzymes in the same pathway to avoid a form of drug resistance involving substrate accumulation, which they named 'metabolic resistance'.[40]

Christopherson moved to the University of Sydney in 1986 as a lecturer in what is now the School of Molecular Bioscience. For the past twenty-eight years, he has worked on the mechanisms of action of anti-cancer drugs, elucidating the antipurine mechanism of the widely used antifolate, methotrexate and the multiple mechanisms of fludarabine. In 1998 he developed a CD antibody microarray called DotScan for diagnosis of leukaemias based solely on an extensive profile of surface proteins. DotScan has since been used to obtain extensive immunophenotypes of other cancers such as colorectal cancer and melanoma. More recently, 'engineered antibodies' called demibodies have been designed that will selectively kill only cancer cells that express an unusual pair of surface proteins (CD antigens). In 2003, Christopherson obtained substantial funds for a Major National Research Facility for Cancer Proteomics with mass spectrometers and other equipment required for two-dimensional fluorescence difference gel electrophoresis (DIGE), and isobaric tags for relative and absolute quantitation (iTRAQ) coupled to two-dimensional liquid chromatography (2DLC) and tandem mass spectrometry (MS/MS). His current research continues in the area of cancer proteomics.

Richard Christopherson's work on leukaemia includes investigation of the mechanism of action of drugs such as cladribine and fludarabine against chronic lymphatic leukaemia and hairy cell leukaemia by using DotScan microarrays and DIGE; surface profiling of colorectal cancers and detection of tumour-infiltrating lymphocytes; and classification of leukaemias by cell surface profiling.

The career of Richard Noel Pau (1938–2004) was cut suddenly short; the last paper he co-authored, dedicated by his collaborators to his memory, was published the year after his death. Pau came to the University of Melbourne when his wife, Jaynie Anderson, was appointed to the Herald Chair of Fine Arts in 1997. Pau became a principal fellow with the title of Associate Professor in the School of Chemistry as well as the Department of Biochemistry and Molecular Biology. He soon established himself as a valued collaborator and postgraduate supervisor, notably after the sudden death of Barrie Davidson, whose PhD supervision he assumed. Richard Pau also published several papers with colleagues from Melbourne. Both his co-authored paper on the biochemical and thermodynamic characterisation of the molybdite binding protein Mop from *Haemophilus influenzae* and his chapter on molybdenum uptake and homeostasis in *Genetics and Regulation of Nitrogen Fixation in Free-Living Bacteria* were published after his death.[41]

The career of David Lee Ebert (1959–) took a trajectory that runs counter to that of many university staff. Many academics, including biochemists, have turned from secondary school teaching to university teaching and research. Ebert went in the opposite direction. He summed up the reaction of his colleagues in an article in *Australian Biochemist* at the end of 2004:

'You're doing what?!?'

This was a not uncommon response when I began telling people that, after ten years, I was resigning my position in the Department of Biochemistry and Molecular Biology at the University of Melbourne to become a primary or secondary school teacher. Why would I do such a thing? Did I not enjoy research? Was I bored with lecturing?

Was I disenchanted with the department or the university? The answer to each of these questions is a resounding no, no and no![42]

Ebert, who took his BSc (Hons) in 1981 from the University of California, Davis and his PhD from the University of Wisconsin, Madison in 1986, where he held a Peterson's Postgraduate Fellowship, came to the University of Melbourne in 1993. He had spent 1987 to 1989 as a post-doctoral fellow in the Department of Biochemistry at his alma mater with Alan D. Attie, the Jack Gorski Professor of Biochemistry, holding an American Heart Association Postdoctoral Fellowship in the first year and a Postdoctoral Fellowship from the National Institutes of Health in the second. He came to Australia in 1989 as a visiting scientist in the Lipoprotein Molecular Biology Laboratory of the Baker Medical Research Institute, where he worked with Alana Mitchell until 1992.

David Ebert (*left*) in 2001 with his first PhD student, Craig Clements (now at Monash University) at his graduation

Although he had virtually no teaching experience until he took up a lectureship in the department, and assessed his own initial performance harshly, David Ebert by 2002 was one of three recipients of an Excellence in Teaching Award based on a ballot in which students in the first three years of the medical course nominated their best teachers.[43]

The Ebert laboratory in Biochemistry concentrated its investigations on two proteins that modified lipoproteins in the bloodstream: hepatic lipase and cholesteryl ester transfer protein (CETP). CETP was especially interesting, since although rats appeared not to have any activity, according to one publication, they had the gene and produced CETP mRNA. In Ebert's own words:

> This was of particular interest because at the time it appeared that expressions of high levels of CETP were associated with increased risk for heart disease. The problem was, try as we might, we could not find any evidence of CETP expression and, if there was a gene, it was only a fragment of it. Although I believed the negative evidence of my first Honours student, Alison Roy, it was hard to use it to rebut the published findings. This project sat for nearly ten years until the completion of the mouse genome and partial completion of the rat gave us another angle.[44]

A large number of biomedical science students subsequently trawled through the Celera mouse genome database for any evidence of a CETP gene as part of a bioinformatics class project. Ebert's last honours student, Cathryn Hogarth, spent a summer piecing together the mouse and rat sequences to discover that both the mouse and the rat only had remnants of a CETP gene left and that there was no way a functional gene could be expressed.

By 2002, Ebert had decided that he wanted to change direction and began studying for his Bachelor of Teaching (Primary and Secondary). Although he maintained his association with the department as a fellow from 2005 to 2009, he embarked on a new teaching career, most recently at Malvern Central School, before moving, in 2013, to the position of Head of Science at Glen Eira College. It is fitting that this summary of the influence and effect of great teachers should end with that of a man who decided to start teaching students at secondary rather than tertiary level, among them perhaps some future scientists of note.

There have been many other men and women who, after postgraduate studies in the Department of Biochemistry and Molecular Biology at Melbourne University, undertook further important work elsewhere.

Russell Howard, a pioneer in molecular parasitology, was the overall winner of the Advance Global Australian Award in 2013. His career in life sciences and biotechnology has been dedicated to the generation of products that provide solutions to problems in medicine, agriculture and the chemicals manufacturing business. He spent over two decades studying infectious diseases, primarily the molecular basis for the pathology of malaria and immune evasion by antigenic variation, as well as serving on WHO and USAID advisory panels for malaria vaccine development. At the National Institutes of Health in Maryland, he was responsible for the invention of a rapid, inexpensive, human malaria diagnostic test marketed worldwide. He has established and/or served as CEO of several companies, notably Oakbio, a biotechnology company based in California, and Neuclone, a Sydney company developing biosimilar monoclonal antibody drugs.

Richard Simpson is renowned for his work in proteomics and is another recipient of a Centenary Medal. Working most recently at La Trobe University, his research focuses on the use of an integrated proteomic/genomic strategy directed towards understanding the role of the extracellular environment (specifically membrane vesicles; exosomes) in cancer progression. Various in vitro and in vivo cancer models are used as well as techniques including lipophilic labelling, cell sorting, western immunoblotting, mass spectrometry-based protein profiling for discovery and targeted strategies, and miR/mRNA profiling and qRT-PCR validation.

Nicos Nicola is another Centenary Medal recipient who also won the Gottschalk Medal in 1986, the Wellcome Australia Medal in 1996 and the Amgen Australia Prize in 1996. Since 1997 he has been Research Professor of Molecular Haematology at the University of Melbourne. His laboratory focuses on the molecular regulation of haemopoietic cell production and function. Work on the cytokines that regulate the production and functional activation of granulocytes and macrophages, two types of white blood cells that coordinate innate immune responses to bacterial and viral infections, has led to the identification, purification and/or molecular cloning of several important cytokines and the identification of a new family of inducible, intracellular inhibitors of these cytokine/receptor signalling pathways. Also under investigation is the role of cytokines and cytokine signalling pathways in the development and maintenance of leukaemic cell populations and the usefulness of cytokines or cytokine antagonists as therapies or adjunct therapies in cancer treatments.

Francis Carbone took a position as professor in the Department of Microbiology and Immunology at the University of Melbourne. He previously held appointments of assistant professor at the Scripps Research Institute in California and associated professor at Monash University. On his election as a fellow of the Australian Academy of Science in 2010, the citation noted that he had made a number of critical discoveries about the nature of immunity, specifically defining the function and behaviour of key cells involved in the response against infection, and had identified mechanisms by which the immune system recognises pathogens and how effectively immunity is generated to control these agents. The versatile range of biological tools for the study of immune components that he developed has proven indispensable to the field and are used worldwide.

Another of the department's graduates to take her skills overseas is Yvonne Paterson, Professor of Microbiology at the University of Pennsylvania. Paterson took her PhD from Melbourne in 1978 on the effect of side chain complexity on the dimensions of polypeptide chains. Her research currently focuses on rational approaches to immune intervention in neoplastic and infectious disease.

Robin Anderson, who holds honorary positions in the Department of Pathology and the Sir Peter MacCallum Department of Oncology, took her BSc from the University of Melbourne and her PhD from La Trobe. Her laboratory aims to understand the molecular events underlying the spread of cancer cells to other parts of the body, hoping to develop new approaches for the treatment of cancer that has metastasised.

The career of Robin Anders, who, like many Melbourne graduates, found a final home in the Department of Biochemistry at La Trobe University, took a rather different path. Following his graduation in Agricultural Science at Melbourne in 1965, he worked in Papua New Guinea before returning to his alma mater for his PhD studies.[45] He spent 1965 to 1974 in Papua New Guinea and then returned to Australia. He was chairman of the organising committee of the Lorne Protein Structure and Function Conference for fifteen years.

The final two people in this long list both spent time in the Department of Biochemistry and Molecular Biology after taking their doctorates there. The laboratory of Alan James Hillier, who took his PhD on aspects of the metabolism of carbon dioxide by *Streptococcus lactis* in 1976, was physically located within the Russell Grimwade Building from his appointment in the CSIRO Division of Food Research until he moved his laboratory to Werribee. The move, in 1998, occurred when Hillier, who headed the Cheese Science and Microbial Biotechnology Section of what was then known as Food Science Australia, took a half-time appointment as professorial fellow with the Department of Food Science of the newly established Institute of Land and Food Resources of the university, while retaining a half-time appointment in Food Science Australia.

Hillier's career-long focus on the biochemistry, physiology and molecular genetics of lactic bacteria initially involved studies on their metabolism. This changed with the advent of recombinant DNA techniques and the new knowledge of the physiology and metabolism of lactic acid bacteria which they facilitated. His research, at the interface of academic research and its application to industry, has involved cooperation with Dairy Australia and the country's major cheese-producing companies as well as with research institutions. He has received numerous professional awards, including the International Dairy Science Award of the American Dairy Science Association in 1999.

Melissa Anne Brown completed her PhD studies in the Cancer Research Unit of the WEHI in 1993.[46] Her postdoctoral training at the Imperial Cancer Research Fund and King's College in London contributed to the characterisation of the first breast cancer susceptibility gene, BRCA1. She lectured in the department between 1997 and 1999, during which time she won the Selby Research Award. She left Melbourne in 2000 for the University of Queensland, where she is currently Professor, Research Group Leader and Deputy Dean of the Faculty of Medicine and Biomedical Sciences. Her research, funded by the NHMRC, the National Breast Cancer Foundation, the Queensland Cancer Fund and the University of Queensland, is focused on understanding the regulation of breast-cancer-associated genes. Brown won a University of Queensland Foundation Research Excellence Award in 2002, an International Union against Cancer Technology Transfer Fellowship in 2005 and a University of Queensland Award for Excellence in Research Higher Degree Supervision in 2010.

Melissa Brown, Professor, Research Group Leader and Deputy Dean of the Faculty of Medicine and Biomedical Sciences at the University of Queensland

The University of Melbourne takes enormous pride in the success of its alumni. While their achievements are due principally to their own talent, determination and the stimulus and support of the organisations and groups with which they were later associated, it is with pleasure that their alma mater records that they got their start in Biochemistry at Melbourne, along with the hope that they are the forerunners of other scientific stars.

The Gleeson Laboratory in the Russell Grimwade Building

CHAPTER 8

'The Academics Couldn't Function without Them'

T HE REMARK THAT heads this chapter was made by Valda McRae (1935–2014), a scientist with a long history in administration in science at the university. It could be echoed by virtually all her colleagues, referring to the vital function of the technical staff working in the laboratories and workshops, staff such as Archibald Gallacher, John Ford, Archibald McEachern and Elizabeth Minasian who were all held in high regard. Many of these men and women spent several decades in the department and the division between technical and academic personnel is not as clear-cut as it might at first appear to be. Some, such as Elizabeth Minasian, had professional qualifications in a field related to biochemistry; others held postgraduate academic qualifications in science.

In 1975, for example, the technical division of the department was divided into five main sections, under the general direction of a laboratory manager: Ancillary Services, which included supply, stores and transport, solvents and cleaning; Animal Rooms; Teaching Services; Research and Postgraduate Services; and Equipment Services, which was itself divided into Electronics and Mechanical and General. The postgraduate and research laboratories were essentially run by the various academic staff, with technical assistance from people like Elizabeth Minasian, but non-academic staff had direct responsibility for the administration of the teaching laboratories.

One experience shared by academic and non-academic staff in the decade or so after World War II that may have contributed to the mutual regard in which they held each other was that of war service. Academic staff who had witnessed firsthand the capacities and strengths of the mechanics and technicians serving with them in the military would have been equally ready to acknowledge these strengths in civilian life. The war in fact brought to the university many men who had acquired skills as well as the confidence to use them innovatively.

One such man was Ian Still Young (1924–2009), who joined the department in 1955 as a technical assistant grade 2 (reclassified as grade 3 the following year) and retired more than three decades later in 1989 from the position of technical officer grade 1. Young had spent his early years in Charters Towers and his war service (he was living in Kelvin Grove, Queensland at the time) was undertaken with the 19th Field Ambulance Unit in the Solomon Islands and in Bougainville, New Guinea.

Young's position was that of the department's animal house technician and he also took responsibility for ensuring that the Biochemistry building was locked at night. After he had been with the department for twenty years, Head of Department Kai Mauritzen noted:

> The production and maintenance of healthy animal stocks is an essential prerequisite for success both in research and practical classes. It is not too much to say that, for us, the health, quality and reproducibility of our animal supply ranks in importance with the supply of chemicals. For these reasons our Animal Section must be classed as a vital service facility.
>
> Mr Young is among the most devoted and loyal members of staff. He works long hours over weekends and Public Holidays without complaint to ensure the well being of the animals in his charge.[1]

This assessment was echoed by Gerhard Schreiber, the Professor of Biochemistry (Medical), who supported Young's reclassification as a technical officer grade 1 and gave an indication of his duties:

Our research group is one of the major users of the animal house facilities of the department. Among other projects, research is carried out on minimal deviation heptoma which grow only in special inbred strains of rats. Mr Young's work was essential in rapid establishment of the two required inbred rat strains. It is of great importance for our experiments on the regulation of protein synthesis in heptomas and normal rat liver that the animals are in a comparable nutritional state. This can be achieved only by a carefully controlled feeding schedule. Animals also have to be absolutely free of all infections. These conditions are not met in commercially available rats. We were very impressed by the high standard and quality of the rats bred by Mr Young. Any department involved with animals is very lucky if it can keep an animal house officer of the skilfulness, dedication to his work and reliability comparable to that of Mr Young.[2]

The payment and classification of the non-academic staff, however highly skilled, were frequent sources of tension in many departments as they fought to retain the services of people with sought-after skills. In 1954, shortly before Ian Young's arrival at Melbourne, the University Council was asked to remedy a 'serious difficulty over many years' which had 'arisen from the grading of a small group of highly expert and responsible senior technicians as Laboratory Stewards with the salaries of lecturers'.[3] Later council decisions had restricted the grading of lecturer to academic positions, and these extremely competent and greatly respected members of staff had found themselves being paid no more than the people they were supervising.

When the council was asked to look at the situation, it affected only five men in the university and the anomaly was resolved by promoting them from the top of their previous salary range (grade 7) to the bottom of a new classification (grade 8), which gave a laboratory steward a salary of between £1150 and £1300, compared with the starting salary of a lecturer which was £1550. Under the new classifications, laboratory technicians and workshop personnel were paid according to the Metal Trades Award, whereas building maintenance staff came under the Building Trades Award. All female technical assistants were paid £40 less than men doing the same work. It is instructive to compare the pay scales of other skilled non-academic staff with the rates paid to lecturers. Patternmakers and scientific instrument makers received £783, fitters and turners who were machinists first class got £749, machinists second class got £710 and machinists third class were paid £686. A carpenter's salary was £841 and that of a labourer £723, while painters and plumbers earned £835 a year. Senior tutors and senior demonstrators were paid between £900 and £1500.

Considerable discussion ensued over what to call the five recently elevated men. On 24 July 1956, the Staff and Establishments Committee noted (without, sadly, recording how much time they had spent on this exercise) that 'being unable to suggest a better name for this category of technicians', they had 'decided to let the matter lapse'. In 1956, when Joyce Calvert was appointed, the salary of a female cleaner was £11.1.8 a week, or £576.6.8 a year.

Poor salaries were not the only reason for staff departures. Although Archibald McEachern had resigned his position in charge of the department's workshop in 1949 for financial reasons, leaving for work in the Kiewa Valley because Trikojus had been unable to persuade the university to increase his salary to an adequate level, Monash University, which opened its doors to students in 1961, and La Trobe University, which took its first students in 1967, proved just as attractive to many technical staff as they did to their academic colleagues, with Biochemistry among many departments affected. Many people cited ease of getting to work rather than financial reward as a prime reason for leaving.

Trikojus's long-serving and much-valued laboratory manager, Archibald Gallacher (another war veteran), was by no means the only member of the department's technical staff to take up a position as a foundation member of the La Trobe University School of Physical Sciences. John Chippindall (1911–1993) spent less than four years at the University of Melbourne, but by the time he left, he had made a notable mark. There are, however, some interesting discrepancies in his life history before that. According to his Australian Army record, John Chippindall was born in 1911 and left AHQ Signals with the rank of captain; according to the university's records, he was born in 1913 and held the rank of major.[4] Furthermore, his letter of application to Gallacher lists a work history which makes clear the range and depth of his working life but is remarkable for its failure to identify specific institutions and dates:

Details of Experience:

Subsequent to initial training, a varied activity in general workshop functions engaged on precision engineering, laboratory equipment, special manufacturing plant and general jobbing work, the last nine years in the senior position as Chief Engineer.

During the war years, service with the A.I.F. with the rank of Major, responsible for the design and manufacture also operation of long range radio transmitters prior to Industry being geared to supply.

Presently engaged in Electronics, specialised airborne navigational Doppler Radar, Tonsight Radar, Computers, Electronic Equipment, Communication Receivers and other Electronic test gear, U.H.F. and micro-wave Equipment, etc., for the R.A.A.F., R.A.F. and Department of Supply, also Marine Radar, Depth Sounding Equipment of advanced design for shipping and general purposes.

Some 20 years active operation of an amateur radio experimental station as a means of developing further knowledge of the subject.[5]

Without specifying the institutions involved (although the university must surely have asked for details), he described his educational qualifications as:

Basic Workshop training as Indentured Engineer supplemented by Technical School training to Diploma Standard.

Five years Marconi training in advanced Electronics.[6]

Chippindall was appointed in April 1963 at the top of the salary range for a senior technical officer because of his age and experience, and his work so impressed Trikojus that just two years later the professor wrote to the Deputy Vice-Chancellor urging Chippindall's promotion to chief technical officer:

I think he must be regarded as one of the best-qualified specialists in electronics on the technical staff in the University. He has made himself familiar with all the equipment of a major as well as a minor character in the Department, with electrical or electronic components. Recently he was invited to assess in Sydney the relative merits of three commercially available liquid scintillation spectrometers, at the expense of the companies concerned. As a result, the company from which we intend to order such equipment has agreed with his recommendations for important alterations in design. He is highly original in his approach to equipment and has incidentally produced a revolutionary new design for the Warburg bath used for manometry and has manufactured a number of them in the Department, including one for the School of Botany.[7]

John Chippindall in Chemistry with a work experience student from Macleod High School, October 1974

Not even the promotion he was awarded in 1965 was sufficient to retain his services, however, and in October 1966 he wrote to the professor:

I have for some time followed with interest the development of La Trobe University and considered carefully the effects that transfer of my services to this new university would have upon my personal carcer and the general problems associated with access to and from employment. The establishment of such an undertaking offers an interesting challenge both technically and physically, and, from my home, freer access in regard to travel and traffic problems.

As result I have decided to accept appointment to La Trobe University as Chief Technician Workshops in the School of Physical Sciences. Therefore, I reluctantly tender my resignation as Chief Technical Officer in the School of Biochemistry, effective from November 2nd, 1966.

Please be assured that this serious step has only been taken after prolonged consideration and it is with regret that I terminate an inter-esting and happy association. May I also thank you personally for your guidance and understanding during the period I have been privileged to work in your Department.[8]

John Chippindall's career at La Trobe University, where he stayed for the rest of his working life, was no less distinguished. Under the direction of Professor J.D. Morrison, his specific task was to build mass spectrometers. One of his colleagues recalls that 'Mr Chipps' was most famous for the construction of the 'world's only' Field Desorption Mass Spectrometer, built for Peter Derrick's research. This was something like 20 feet long and 10 feet wide, completely filling the room in which it was located. There was, according to people who worked there at the time, some resentment among other staff members because their own workshop projects were continuously delayed by this endeavour.[9] This 'grand-scale magnetic-sector mass spectrometer that featured an 8-ton magnet' followed Derrick to the University of New South Wales and eventually to the University of Warwick.[10]

Another war veteran was Francis Charles Baker (1924–2007), who joined the department in 1966. Baker had served in the RAAF with the rank of flying officer from 1943 until the end of 1945. Having obtained his Associate Diploma in Applied Chemistry, Associate Diploma in Chemical Engineering and Post Diploma in Industrial Management, from 1949 to 1962 he worked as a chemist in the Chief Engineer's Branch of the Postmaster General's Department, the Albion Explosives Factory and the Mulwala Explosives Factory. On moving to the Altona Petrochemical Company owned by Esso-Mobil, he supervised the Operating Department. As shift superintendent he oversaw all direct production activities;

Francis Baker, laboratory manager

in the capacity of area supervisor he was responsible for planning and applying budgetary controls, and planning maintenance schedules for the process and steam raising plant. He also trained operational and supervisory staff and was responsible for safety matters.

As laboratory manager in the Department of Biochemistry, Baker's responsibilities were described in 1971 as encompassing the management of day-to-day technical purchasing and accounting operations in the routine running of the department, supervision of a technical and non-academic staff of thirty-three people, recruiting them and allocating their duties, training junior technical staff, maintaining continuous review of safe working practices, overseeing the departmental budgetary situation and verifying all trade invoices, reconciling more complex purchase orders and resolving supply queries and disputes. He was also expected to maintain and install all departmental scientific assets to their optimum specifications, critically evaluate new or additional equipment and liaise with the Maintenance Department and outside contractors as well as providing janitorial services.

The control of consumables involved responsibility for safe and secure storage of materials, bearing in mind issues of flammability, noxious or radioactive hazards, sterility requirements, prevention of deterioration, and contamination or spoilage of valuable and/or labile biochemicals. Baker was also responsible for the departmental animal rooms, covering their maintenance, staffing and general efficient operation. As we have seen, this involved oversight of general animal husbandry, humane handling, adequate accommodation and post-operative veterinary nursing.

Baker's other duties included liaison with outside bodies such as the Department of Health regarding import controls, the Commonwealth X-Ray and Radium Laboratory on matters relating to radioactive materials, and various divisions of CSIRO as well as the Animal Technicians' Association. Ensuring that the treatment plants coped with the disposal of effluents to the standards of the Melbourne and Metropolitan Board of Works and that the incineration of animal carcasses and disposal of low radioactivity wastes and solid toxic materials adhered to Health Authority standards, formed another part of his duties. He also ensured that steam or radiation sterilisation facilities were available for equipment. Finally, he was responsible for liaison with the central university authorities to ensure security of department personnel, documents, student records and cash.

Frank Baker left the department in 1973 to take a position as chemist class 3 in the Commonwealth Public Service. His successor, Neil James Griffiths (1942–), also spent only seven years in the department, leaving in 1980 to take up a position at the Explosives Factory, Maribyrnong. Griffiths held a Diploma of Applied Chemistry from RMIT and spent 1959 to 1966 in the Explosives and Ammunitions Section of the Defence Standards Laboratories in Maribyrnong, where he worked initially on the analysis of explosives and their constituent raw materials, and later in a position that had more in common with fitting and turning or mechanical engineering than chemistry, as part of a team developing bomb and mortar fuses. From 1966 to 1973 he worked at RMIT in the Organic Section of the Chemistry Department, where he was directly responsible to the senior lecturer in charge of Organic Chemistry.

As laboratory manager grade 4 (as his position was reclassified in May 1974), Neil Griffiths was responsible to the Head of Department for all technical activities and their overall administration in a large department involving teaching and research. He was expected to have a broad understanding of general university administrative procedure, with particular emphasis on buildings and maintenance accounting and staffing functions. His responsibilities included the selection, training and supervision of technical staff, both teaching and research, and participating in special projects by giving practical assistance and advice to members of the academic staff, postgraduate staff and students. His areas of control included workshops, photography, animal breeding and welfare, electron microscope operation and maintenance, multidiscipline teaching laboratories, cleaning and portering, safety, purchasing and general budgetary and financial control of all related areas.

Neil Griffiths, laboratory manager, in 1974 with his daughter, Laura Jane

Griffiths also held the position of safety officer. A 1977 memorandum from him to all staff and students illustrates some of the hazards involved, in the form of a mishap the daily press reported as 'The Vile Vial Saga':

Our-Faces-Are-Red-Department
Re: Handling of Toxic Materials

A recent incident occurred in the Department in which a vial of a very toxic material was thought to be lost somewhere between the Biochemistry and Agriculture Departments. As it turned out, the chemical <u>never left</u> the basement Store even though internal delivery records indicated that it had been delivered to a laboratory in the Department and, indeed, signed for by a person in that laboratory! In the meantime, extensive searches were undertaken and the police called in to assist.

This example highlights the need to be extremely careful when dealing with toxic compounds. It is the responsibility of individual members of the Department to make sure that:

1   They are fully aware of the toxicity of all reagents and chemicals used by them.
2   They use appropriate safety procedures for handling or disposal of all such toxic or dangerous materials.

3 They instigate proper check procedures to monitor incomings and outgoings of all such toxic materials.
4 Highly toxic compounds are kept under lock and key in the laboratory's poisons cupboard.
5 When toxic materials are to be transported outside the Department, an appropriate double container technique is used to prevent the possible loss of any of the material.

Fortunately, this time we were lucky, but we should treat this example as a warning of what might have happened.[11]

Neil Griffiths left the department in 1980 to take up a possibly less stressful position at the Explosives Factory in Maribyrnong running the nitroglycerine manufacturing facility.

Two men who joined the department in 1962 and worked there until the 1990s were Robert Goullet and Graeme Strange.

Robert Edward Goullet (1931–2009), who, with Albert Fairchild, succeeded to part of John Chippindall's position, taking responsibility for the electronics component of the equipment services in the department, came back to the university after eleven years with employers as varied as MacRobertsons Ltd, the Government Ammunition Factory at Footscray and the Department of the Navy. He had spent 1954 in the university's Department of Mechanical Engineering workshop. Goullet was appointed as a technical assistant grade 4 but a mere twelve months later was promoted to the top of the range for technical officer grade 1. At the beginning of 1967, the staff officer told the Deputy Vice-Chancellor:

> His basic training was as a Fitter & Turner, and he has since developed skills in other fields. He is a highly skilled machinist, used to working to fine tolerances and able to use a great variety of workshop machine tools.
>
> He has recently qualified as a Radio Technician, and has developed the ability to construct and repair electronic apparatus and instruments …
>
> In analysis, Mr Goullet is a skilled and experienced all-round technician … Mr Goullet has beyond doubt proved his technical attainments. He has also evidenced his willingness to accept additional

loads whenever required of him, and his ability to efficiently carry them out. He is of great assistance to Mr Fairchild at present while that officer is adapting to the particular needs of the Department.[12]

Less than ten years later, the Head of Department, writing in support of Goullet's promotion to chief technical officer, which he achieved in 1976, gave a clear picture of the man and his duties:

> Mr Goullet is responsible for the maintenance and repair of a large range of electrical and electronic equipment in the Department. His work includes preventive maintenance of the equipment and, where breakage or defects are reported, involves examination and diagnosis of the trouble followed by repair or rectification of the fault. In addition to repairing and maintaining equipment he also undertakes the fabrication of new instruments. Currently, in association with Dr W.H. Sawyer, he is building a fluorescence polarization instrument. Such equipment is not commercially manufactured and, when built, will be the first of its kind in the southern hemisphere.
>
> This Department is totally dependent both in research and teaching on the operational maintenance of a large range of electronic equipment. The total replacement value of this equipment is of the order of $480 000; I can supply a detailed list of the equipment serviced by Mr Goullet if required. Without Mr Goullet or someone like him to service our equipment, the Department would suffer a rapid rundown in efficiency … Mr Goullet's attitude to work is unusual. He derives genuine pleasure from helping the academic staff with instrument problems and he delights in the constant technical challenge presented to him by his work. For these reasons he works long hours, including weekends, and seeks no remuneration (indeed he refuses it) for the additional time he spends outside normal working hours.[13]

By this time, Robert Goullet had taken various mechanical and electronic engineering subjects at the technical college level, attended the course on medical and biological electronics held in the Department of Physiology, and possessed a full PMG licence as an amateur radio operator. He was an active member of the Wireless Institute of Australia and the Western and Northern Suburbs Amateur Radio Club, which noted in its obituary tribute his 'ability to unearth information on a particular rig, amateur radio program, operating or service manual' which he would then distribute, without any thought of recompense, on CD or print to whoever wanted it.[14]

By 1990 Goullet had agreed to assume responsibility for supervising the installation of an extensive PC-computer-based LAN within the department in addition to his duties as officer in charge of the department's Electronic Workshop. The Head of Department succeeded in having his position reclassified as chief technical officer grade 2, noting:

> Mr Goullet has considerable computer expertise (self-taught), both in hardware and software, and has already assumed a pivotal role in the computerisation of the Department. While this role has been absorbed into his present job specification, Mr Goullet's contributions have been well beyond the call of duty, and his enthusiasm and considerable skill in the computer area has been a major driving force in our recent development as a major PC user within the University and an innovator in the area of computer-aided instruction (Biochemistry field). In addition, his intricate knowledge of PC hardware and software has resulted in considerable savings to the Department, particularly in the area of purchasing and maintenance … If the request is not granted, there is a real possibility that Mr Goullet will resign to seek further opportunities in the PC industry, which would be a disaster for the Department.[15]

Robert Goullet retired from his new position in 1992.

Robert Goullet (*left*) with Albert Fairchild, John Morley and Lindsay Rayner in 2001, on the occasion of John Morley's retirement

Graeme Edmund Strange (1936–2001) did not leave the University of Melbourne to take up another position either, although Trikojus expressed grave fears that he might do so. Strange came to the department in 1962 as a technical officer grade 2 in charge of stores. His previous laboratory experience had included work in cereal chemistry with Bunge (Australia) and W.S. Kimpton and Sons. Work in the laboratories of Commonwealth Fertilizers, a division of the Imperial Chemical Industries

Graeme Strange, technical officer in charge of stores in 1973

of Australia and New Zealand, had given him experience in general analytical chemistry. In 1967, Trikojus recommended him for promotion to senior technical officer grade 1, warning of dire consequences if this recommendation were unsuccessful:

> He has revolutionised our storekeeping and received very favour-able comments from the Parliamentary committee which visited the University earlier this year. He is currently undertaking an advanced course in Purchasing and Supply Management at the Royal Melbourne Institute of Technology. He is also a skilled photographer and handles a proportion of the departmental slide requirements. He is also responsible for solvent recovery and purification processing. It would be a critical loss to the Department if he were to leave. We have already lost a total of four key people on the technical side to Monash and La Trobe.[16]

In August 1970, *Staff News* reported an event which had already drawn congratulatory letters from the Vice-Chancellor to Strange and another member of the Biochemistry staff, A.C. Fairchild:

> Prompt action by two members of the technical staff of the Russell Grimwade School of Biochemistry prevented major damage being caused by a major fire.
>
> Senior Technical Officer Mr G.E. Strange quickly raised the alarm when he noticed a small fire in the School's Solvent Distillation

Laboratory and later he helped to extinguish the fire. At considerable personal risk, Chief Technical Officer Mr A.C. Fairchild went into the solvent storage room and extinguished the fire. As the result of the fire some paint on the walls and ceiling of the laboratory was smoke-affected and venetian blinds and a light fitting were damaged. Without the prompt and effective action taken by Mr Strange and Mr Fairchild, the whole building could have been destroyed.[17]

Having taken his Diploma in Purchasing and Supply Management in 1969, Strange completed the Certificate of 16mm Motion Picture Operator at RMIT in 1973. The previous year he had been promoted to senior technical officer grade 2, with Francis Hird providing a note that allows greater understanding both of the man and of the duties of the position:

> During his University service he has developed the Chemical and Apparatus Stores of the School of Biochemistry to meet the growth in student numbers and the increase in research activity over that period of time. He has extended the range of sources of supply of technical materials consumed by the Department, in particular those requirements which are exclusively biochemical in nature. Other technical services under his supervision include the purification, pre-treatment and recovery of solvents used in the Department, this latter activity providing considerable savings within the apparatus budgets of recent years.
>
> Within the Technical Staff organization of the School, the position occupied by Mr Strange is under the direct supervision of the Laboratory Manager. Two Technical Assistants are directly supervised and provide support in routine, manipulative areas of the work.
>
> It is proposed to delegate responsibility for:
>
> 1 Servicing the building by day staff and contract cleaning labor. This will add a Porter and three or more Cleaners under the direction of this position and will also require close liaison with the Services Section of the Buildings Branch.
> 2 Organizing, scheduling and care and maintenance of equipment used in supplementary audio-visual teaching techniques and training of non-academic staff.[18]

Graeme Strange retired in 1990, after almost thirty years in the department. In 1975, as the senior technical officer in charge of Ancillary Services, he was responsible for the staff of four cleaners. Robert Francis O'Reilly (1913–), originally designated a porter, was one. The history of some of these men is recorded in *Minding the Shop*, and it is notable that many, such as F.R. Dart and the long-serving Leonard Munro, were promoted to their positions after being first engaged as laboratory assistants. Porters were initially referred to in University Council papers as messengers and they continued to perform these functions, carrying parcels and so on between university buildings or to and from the city.[19] The equivalent staff are now known as buildings managers and take responsibility for a wide range of tasks, typically in several buildings in a particular zone of the university grounds.[20]

The first porters to service Biochemistry were attached to the Department of Physiology, of which Biochemistry formed a part.[21] At one time, O'Reilly, who joined the department in 1958 and retired in 1978, also supervised its cleaning staff (working under Graeme Strange). He was promoted to porter grade 2 during the 1970s (the staff records are far from clear about the date) and at his retirement was designated a cleaner, possibly in recognition of his supervisory role. The work of porterage is now undertaken by staff whose remit covers several buildings: cleaning services are contracted out. Despite his long service, very little is definitely known about O'Reilly. He was born in Belfast, emigrated from Ireland to Australia and suffered from a knee injury which may have been the result of involvement with the Irish Republican Army.

Albert Cyril Fairchild (1931–), with whom Strange fought the potentially catastrophic fire in the Russell Grimwade Building, for which they were both commended personally by the Vice-Chancellor and praised in *Staff News*, spent thirty years at the university, twenty-nine of them in Biochemistry. Educated at Preston Technical College, Collingwood Technical College and the Melbourne Technical College, where he qualified as a fitter and turner, Fairchild's first job was as an instrument maker in the research laboratory of the Gas and Fuel Corporation. In 1955 he married and moved briefly to Gippsland to work on the dairy farm owned by his wife's family. Having decided that rural life was not for them, the Fairchilds returned to Melbourne where Bert went to work for the Bureau of Meteorology: he was in charge of the instruments section for the next seven years. In the early 1960s he worked in the Time Systems Division of IBM and in March 1966 he came to the university, spending nine months as a senior technical officer grade 1 in the Physics Department before transferring to Biochemistry in December as the officer in charge of the departmental workshop, taking over John Chippindall's supervision of the mechanical and general aspects of Biochemistry's Equipment Services.

In 1968 Fairchild was promoted to the rank of chief technical officer, which previously had been held by Chippindall. In 1993 his position was reclassified to higher education worker (HEW) 7 and in 1994 he was awarded a University Medal for distinguished service, principally for something that impacted on many thousands of people outside the department.

For almost three decades, initially working overtime and subsequently on contract after his retirement, Albert Fairchild was responsible not only for restarting the Stevens clock in the tower of the Old Arts building, which is emblematic of the University of Melbourne, but also for maintaining all the master and slave clocks in the university system. The bell of the Old Arts clock was silent from 1964 until Fairchild restarted it in 1979. In his original brief in 1978, he had been asked only to assist in the twice-yearly adjustment necessitated by the introduction of daylight saving to Victoria. He went on to assume responsibility for repairs and maintenance to time switches on all security lights and for the maintenance of all the university's ordinary synchronous clocks. Many of these were so old that parts were no longer available: Fairchild made them himself.[22] The bell of the Stevens clock tolled until the early 2000s, when the sound was briefly considered disturbing. In 2010 the mechanical mechanism was replaced with a digital system, but the clock remained silent. In 2014, however, Ingrams Clock Makers designed, built and installed a new mechanism and the bell tolled once more.

Fairchild's skills were, of course, principally applied to equipment in Biochemistry, with Head of Department Dick Wettenhall noting in 1994: 'Bert's skill and ability at maintaining both new and aging equipment is to be commended. He is an integral part in the smooth running of the Department. His efforts are greatly appreciated.'[23] That same year, Albert

Fairchild was made a Member of the Order of Australia in the Australia Day honours list for his services to the Scout Movement.

John Daryl Morley (1942–) is another staff member who spent a short time in the department, left for another institution and then returned to the University of Melbourne. Morley took his Diploma in Applied Chemistry from Swinburne Technical College in 1963 and a position as technical officer in Biochemistry the following year. From 1965 to 1970, he worked at Monash University under Anthony Linnane, passing several subjects of the BSc before moving to the Medical Research Centre at Prince Henry's Hospital, led by Henry Burger, where he worked under Terry Bellair who was investigating the amino-acid sequence of the protein growth hormone. He returned to the University of Melbourne Department of Biochemistry and Molecular Biology in 1972, initially heading the teaching services laboratories. He replaced Ian Croker, who had left, as Chippindall had done, to take up a position at La Trobe University. His position was that of 'class co-ordinator', although the coor-dination involved was concentrated at the end of the academic year when he compiled a master list from those of the chemicals, glassware and larger equipment required for the follow-ing year that were submitted by the three class laboratories. His day-to-day responsibilities covered helping to develop small-scale experimental procedures and operating an 'auto-matic analyser' as well as setting up the teaching areas. As assistant to the laboratory manager he maintained various lists of lockers, equipment and so on, and prepared diagrams for lecture slides and journal articles; he also assisted in maintaining the department's highly complex finan-cial records at a time when computer records were not yet in use.

John Morley, building manager

When the accounts were finally computerised and the position of departmental accounts officer created, Morley's position was reclassified as building manager. At the celebration of his twenty-five years of service in 1997, when he was awarded a university Bronze Medal, he described a typical day as starting with a timetabling clash in the Trikojus Lecture Theatre, then involving finding a new face-shield for a student, moving furniture to reconfigure a teaching space and dealing with building issues ranging from a power failure in the cold room to water streaming into a laboratory because visiting workmen had cut through a pipe in the plant room above it. He retired in 2001.

This chapter deals only with the most senior of the technical staff of the department in the period roughly between 1950 and 1990. In 1975, for example, there were over three dozen technical assistants, eight technical officers and four laboratory attendants. It is simply not possible to name them all, even though many of them spent several decades at the university. But it is worthwhile to look at one woman at least who spent thirteen years in Biochemistry and whose career resembled that of many others.

Jacqueline Ann Camilleri (1970–) came to the university as a junior technical assistant in 1990 after completing secondary school at Chisholm College, Braybrook and having passed four first-year science subjects at La Trobe University: Biology A (Plants & Animals), Biology B (Genetics & Evolution), Chemistry 1B and Geology 1. The prerequisites for the position, as described in the advertisements in *The Age* in January 1990, specified that applicants should have completed their Higher School Certificate, including Chemistry, and must intend to proceed part-time towards higher qualifications. The duties included assisting in setting up laboratory equipment and the preparation of reagents for practical classes in Biochemistry.

Jacqueline Camilleri noted in her application that none of the holiday jobs she had held were scientific in nature, but listed her recreational interests as marine life and sea studies, fishing and scuba diving, land conservation and biological sciences. Her father, Joseph, a cabinet-maker, had migrated as a young man from Malta after World War II and his only daughter led what she characterises as a sheltered life. Her mother long recalled how scandalised the family was when, at the age of twenty-one,

Jacqueline was escorted by a colleague to Elgin Street, Carlton for lunch at the Clyde Hotel, where she drank her lemon squash and ate her food not in the ladies' lounge, but in the bar.

In 1992 Camilleri began part-time study towards the Associate Diploma of Applied Science at the Western Metropolitan College of Technical and Further Education, a qualification she achieved in 1997, winning the Most Outstanding Student Award in her faculty. Her annual performance assessments showed a steady increase in the responsibilities she assumed, ranging from the running and preparation of classes, to the recognition and management of chemical hazards and chemical handling, contribution to the time management and organisation necessary to meet deadlines, and rendering first aid to students and others after accidents and illness. In 1999, partly in recognition of her diploma and partly because she helped with the training of new staff and had taken on additional duties as required by SafetyMAP, assisting the business manager in audits and compliance issues, she was promoted to HEW 4.[24]

Jacqueline Camilleri, who had married during her time in the department, resigned in 2004. She now lives on 1.25 acres outside Melbourne, where she and her husband keep themselves and many others in fruit,

Jacqueline Camilleri, long-serving technical assistant, now a farmer, cheese- and olive-oil maker

vegetables and meat. They slaughter their own poultry and rabbits, and make cheese from their goats. They also grow olives and press their own oil.

Dorothy Gillan née Wilson (1920–) occupied a position similar to that of Joyce Calvert for over twenty years. Sadly, her employment records, like those of Evelyn Parkhill, were destroyed, but her own recollections give an idea of the work she performed for more than two decades. She left school at the age of sixteen and by 1943 was employed at the Defence Standards Laboratories in Maribyrnong. There, she was employed in photography and some filling of ampoules and the use of other such equipment, and learned about the glassware used in scientific work. She married in 1942 and left work in 1945 before the birth of her first child. In 1962, with her youngest aged eight, Dorothy Gillan answered an advertisement for work in Biochemistry. The hours required, between 9 a.m. and 2 p.m., were admirably suited to someone who wanted to see her children off to school and be home when they arrived in the afternoon.

The work of cleaner and bottle-washer in the department, which Dorothy Gillan performed until she retired in 1985, requires both skill and care, since the absolute cleanliness of glassware is vital to functioning scientific equipment. The work of cleaners like Dorothy Gillan and Joyce Calvert involved dealing with the demands of large laboratories and classes, often involving hundreds of students from the many faculties which included biochemistry in their undergraduate courses.

The positions of tutor and senior tutor are academic positions. In many university departments, postgraduate students are employed as tutors, for the most part remaining in this rank for only a few years before moving up the academic scale to lecturer positions. One woman, however, occupied a position as senior tutor for almost three decades after other employment in the department.

When Beverley Bencina née Cook (1940–) took her position as senior tutor in Biochemistry in 1982, she was already well known in the department. At the function held in 2008 to mark the demolition of the Russell Grimwade Building:

> Beverley recounted those early undergraduate practical classes, where
> the second-year group of 40–50 students performed their experiments

in the Maxwell laboratory under the supervision of Mary McQuillan. She remembered that if the lessons got a little weary, it was always easy to get distracted by watching the hockey games through the windows on the south side of the classroom. In her third year BSc in 1961, there were 24 students in the practical classes, which were held in the Young Laboratory (the teaching laboratory named after the Foundation Professor of Biochemistry) on the ground floor under the supervision of Jack Legge, Bruce Stone and Frank Hird. Beverley recalled the range of practical exercises, including the interminable Warburg manometer measurements, and visiting the animal house on the fourth floor for the rat dietary experiments. As a special treat at the end of the year, the third-year students were invited into the staff tea room for lunch and also taken up to the roof of the building by Trikojus, to see the wonderful views of the city.[25]

Having begun her biochemistry studies in 1960, Bencina took her BSc in 1962 and her MSc in 1964 in the Department of Experimental Medicine, WEHI, on neuraminidase in vertebrate cells. The following year she began work in the Department of Biochemistry as a research assistant, performing studies on myelin basic protein (a project which occupied her until 1969), and worked as a demonstrator from 1966 to 1970. The next two years were spent in the Department of Medicine at Westminster Hospital, London, investigating variations in urinary Na and K levels in patients undergoing treatment with diuretics. In 1973 she returned to Melbourne initially as a part-time demonstrator and later became a senior tutor before, in 1992, working full-time as a senior tutor, a position she held until her departure in 2010.

As well as co-authoring a number of scholarly papers, Bencina took an active part in both the Lorne Protein Structure and Function Conference and the Australian Society of Biochemistry and Molecular Biology. Her long association with the department meant that over the years Bencina was entrusted with an ever-increasing level of responsibility for undergraduate teaching, coordinating subjects and delivering lectures, as well as engaging in the more usual tasks of planning and supervising practical classes.

Beverley Bencina now tutors in Pharmaceutical Chemistry in the Faculty of Pharmacy and Pharmaceutical Sciences at Monash University. She is also a singer who has performed with the Astra Chamber Choir for almost fifty years. Her other hobbies include travel and photography – her pictures of the Russell Grimwade Building, as well as some holiday landscapes, can be seen in the Medical Building.

Brent Smith (1976–) came directly to the University of Melbourne after taking his Bachelor of Applied Science (Applied Biology/Biotechnology) (Hons) from RMIT University in 1997. After working for six months in Pharmacology, he joined Biochemistry in December 1998 as the technical officer responsible for the Advanced Laboratory in the Russell Grimwade Building, and became the senior technical officer in the Trikojus Laboratory in the Medical Centre. Although he has published in his academic discipline, Smith notes that his retirement plans might involve becoming a mechanic and running a bike shop.[26] In the meantime he is heavily involved in various sports, including as a member (and committee member) of the Melbourne University Karate Club (MUKC). After five years of training with the club, Smith received his Black Belt (1st Dan). The MUKC was founded in 1968, which makes it one of the

Beverley Bencina, long-serving senior tutor

Brent Smith, technical officer and sportsman

oldest karate clubs in Australia. It trains in a style based mainly on the Shuri-te (Okinawan)/Shotokan (Japanese) karate style but incorporates aspects of ba gua (a Chinese martial art) and savate (French kick boxing). It describes itself as being less formal and more social than most martial arts clubs.

Lynn My Tran (1988–) took her BSc (Hons) in 2009 and spent the next two years working part-time as a research assistant in Pathology in Janetta Culvenor's laboratory and with Lindsay Rayner in Biochemistry, where she was responsible for the setting up of second-year practical classes, the preparation of laboratory reagents and cleaning up and waste management. Having majored in biochemistry in her third year and spent her honours year in Heung-Chin Cheng's laboratory, returning to the department in 2012 as a technical assistant felt like coming home.

Lynn Tran, technical assistant, who returned to the Medical Building after her Honours year in Heung-Chin Cheng's laboratory

There are also many professional staff who have spent long periods in, and offered distinguished service to, the Department of Biochemistry and Molecular Biology. One man has spent over forty years in its teaching laboratories. Lindsay John Rayner (1946–) was just nineteen years old when he came to the department as a junior technical assistant in 1966. In 1968 he passed Matriculation Chemistry at University High School and was promoted to technical assistant grade 3 the same year, with his supervisor noting: 'He is an extremely promising young man, and following a re-arrangement of functions in the Maxwell Laboratory, has accepted responsibilities above those expected of a Technical Assistant Grade 2'.[27]

Part-time study, which was expected within the department, could be an exhausting enterprise for someone with a demanding job and family responsibilities, but Rayner completed his matriculation in 1969. In 1975 he took his Higher Technician Certificate in Applied Science from Swinburne Technical College, and by 1981, having undertaken some subjects in biochemistry, he was promoted to senior technical officer grade 2. The Head of Department, recommending this advancement, gave some indication of the range of Rayner's responsibilities, and it is notable that the reclassification was a personal one, with the position remaining at a substantive rank of senior technical officer grade 1:

> The main task is preparation of teaching laboratories covering 600
> 1st year students from 4 faculties. The experiments to be set up vary
> in size and complexity and often involve detailed long-term planning
> and preparation of specimens. Rayner has streamlined the system of
> laboratory preparation and improved training of junior members of
> staff and become involved in some research projects.[28]

Lindsay Rayner took his Associate Diploma in Applied Biology in 1986 and BSc (Appl. Sci.) from RMIT University in 1988. In 1992 the Head of Department successfully argued for Rayner's position to be reclassified as chief technical officer. The background information provided in his submission outlined several professional reasons and a personal one. During the previous two years, the numbers of students in the practical classes had greatly increased. New instrumentation had been introduced: the department's UV-vis scanning spectrometers, ultracentrifuge and automated FPLC chromatography system were considerably more complex than the equipment previously operated in the teaching area for which Rayner was responsible. Forty PC-based workstations also had been introduced and Rayner would be expected to take a major responsibility for their performance. It was considered desirable to have a single person coordinating the provision of complex and expensive equipment between various classes, taking responsibility for deploying the relevant staff and, finally, coordinating the ordering of the necessary consumables and equipment.

After the new arrangements had been trialled for twelve months, in seeking to have the position ratified, Dick Wettenhall noted:

> Although not directly relevant to the reclassification process, I wish to record that I regard Mr Rayner to be an outstanding contributor to the Department in every respect. In particular, his skill and dedication in the management and operation of his practical classes have been major reasons for the success of our practical Teaching programmes in recent years.[29]

Lindsay Rayner (*right*), laboratory manager who joined the department in 1966, with John Morley and Jackie Camilleri

This skill and dedication were to be tested further in 2007 when the Russell Grimwade Building was demolished and accommodation had to be found at short notice for the undergraduate lecture rooms and laboratories as well as the teaching staff. The achievements of academic staff, such as Irene Stanley, and of the removalists, Kent Moving & Storage, in ensuring that this mid-semester relocation from the Russell Grimwade Building to hastily reconfigured space in the Medical Centre on Grattan Street went as smoothly as possible, with minimal disruption of classes, will be described later in this book.

Lindsay Rayner was awarded the university's Bronze Medal for twenty-five years' service in 1991.

Having looked at the technical staff in the department, attention can now be directed towards those who, with Lindsay Rayner and Brent Smith, remained based in the Medical Centre on the Parkville campus, teaching the ever-increasing number of undergraduate students taking biochemistry as part of their degrees, leading to later careers in teaching or research in biomedicine, agriculture and science.

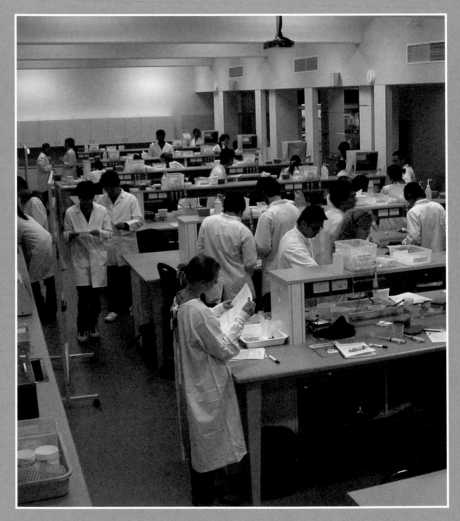

Teaching laboratory in the Medical Building

CHAPTER 9

On the Move Again

AFTER 2005 THE research staff of the department, together with postgraduate students, were relocated in the newly opened David Penington Building on Flemington Road, a part of the Bio21 Cluster, the aim of which is to bring together biomedical research organisations and major hospitals: the University of Melbourne component of this grouping, commonly referred to as the Bio21 Institute, was established in 2002. The teaching staff and undergraduate laboratories remained in the Russell Grimwade Building until its destruction in 2008, when they relocated to the Medical Building.

Undergraduate teaching at the University of Melbourne greatly changed in the first years of the twenty-first century. The Melbourne Model came into effect. This meant that the university stopped enrolling undergraduate students working towards a multiplicity of bachelor degrees, offering instead 'new generation' undergraduate courses. In 2013 there were nine bachelor degrees in Agricultural Science, Arts, Biomedicine, Commerce, Environments, Fine Arts, Music, Oral Health and Science. The teaching staff in the Department of Biochemistry and Molecular Biology thus continued to teach second- and third-year undergraduates from many faculties, including those of Land and Environment, Engineering, Science and Veterinary Science. In 2013 there were five teaching-only staff working in the Medical Building and another staff member who maintained a laboratory presence in Bio21 as well as forming part of this group. Many had a history in the department extending back several decades.

Alana Mitchell (1951–) took her BSc (Hons) in 1972 and her PhD in 1975 with a thesis on the ribonucleotides of *Mycoplasma mycoides*, before working as a senior tutor in the department from 1976 to 1982, during which time she co-authored several research papers with Lloyd Finch.[1] She spent the next three years as a research fellow in the department investigating the regulation of plasma protein synthesis in the rat during acute experimental inflammation. From 1985 to 1987 she investigated the regulation of prolactin gene expression by vitamin D in the Royal Melbourne Hospital Department of Medicine. As a senior research officer at the Baker Medical Research Institute for the next decade, Mitchell chaired the Alfred Hospital/Baker Medical Research Institute Animal Ethics Committee and filled the position of biosafety officer, while at the same time setting up the institute's first molecular biology laboratory and conducting both individual and collaborative research.[2]

Alana Mitchell has a longstanding involvement with science communication and in 1997 worked briefly at the ABC as its science media fellow, preparing material for *The Health Report*, participating in a talkback program on genetic engineering and producing daytime radio on 2BL (now 702 ABC Sydney). In 1998 she established ScienceLink to bridge the gap between scientists and other organisations and individuals – ranging from the general public, stock market analysts and the media to investors and industry – by providing editing and writing assistance. Her interest in animal welfare is demonstrated through her position, from 1998 to 2003, as animal welfare liaison officer with the NHMRC and

Alana Mitchell has long been involved in science communication and animal welfare issues

her current role with the Bureau of Animal Welfare, where she is both a trainer of Animal Ethics Committee members and auditor of Victorian licences to conduct research involving animals. She has also provided assistance on technical writing and editing to the NHMRC Health Advisory Committee. Mitchell returned to a part-time position as senior lecturer in the Department of Biochemistry and Molecular Biology in 1998, combining this with her work as the principal of ScienceLink.

Irene Johanna Stanley arrived in Australia from her native Germany, where she had studied languages and business administration, in 1971. She took her BSc (Hons) from the University of Melbourne in 1979 and her PhD on molecular regulation of granulopoiesis in 1984 at the Cancer Research Unit of the WEHI. Before joining the department, Stanley completed three postdoctoral appointments studying mechanisms of carcinogenesis: she achieved this in 1984–86 as postdoctoral research fellow in the Differentiation Program of the European Molecular Biology Laboratory (EMBL) in Heidelberg, supported by fellowships from the

Irene Stanley is heavily involved in the department's multimedia and online developments and was very involved in the development of the new Melbourne Model curricula

World Health Organization's International Agency for Cancer Research and the European Molecular Biology Organization. Her research at EMBL resulted in one of the first demonstrations that a number of different oncogenes and tumour suppressor genes are required for the progressive development of cancer, with each mutated gene contributing specific properties to the transformed cell.[3] Returning to Australia, she continued working on cancer from 1987 to 1990 as a research scientist in the Colon Tumour Biology Unit of the Ludwig Institute for Cancer Research, and from 1991 to 1993 as senior research scientist in the Department of Cell Biology at the Peter MacCallum Cancer Institute. During this time she also had two research consultancy appointments with the Strategic

Industry Research Foundation and completed a postgraduate Diploma in Information Services at RMIT, for which, in 1990, she won the Student Award of the Australian Library and Information Association which recognises the highest-achieving graduate student of the year.

Stanley joined the department in 1993 as a lecturer, initially employed for 20 per cent of her time as senior research fellow, funded by her own research grants from the Australian Research Council and NHMRC; she was promoted to senior lecturer in 2008. In the mid-1990s the university offered equivalent standing to scientific and educational research, in particular the development of multimedia teaching tools, and Stanley chose to dedicate herself to this. In 1996 she was one of the first recipients of a Committee for the Advancement of University Teaching grant, which she used to develop a multimedia teaching tool, 'The Science of Colon Cancer', that was used (with revisions) in the medical curriculum from 1997 to 2011. Stanley initially taught medical, dentistry and optometry students, lecturing on molecular biology and molecular events in carcinogenesis, while also coordinating and teaching practical courses for which she developed the practical classes 'glucose tolerance test', 'biology of colon cancer' and the 'liver function test' in collaboration with Dr Ken Sikaris, her former biochemistry prac partner; Sikaris is now an associate in biochemistry and the director of Chemical Pathology at Melbourne Pathology. Following the development of the postgraduate medical curriculum in 2011, Stanley's cancer lectures concentrated on molecular disruptions of the cell cycle in the context of haemopoietic neoplasms.

Stanley contributed greatly to the multimedia and online developments of the department, and when she took on the coordination of the second-year Biochemistry and Molecular Biology subjects for the BSc course in 1999–2000, she was keen to develop online support for her subjects. As 'academic developer', she achieved a productive collaboration with the university's website providers, programming multimedia tutorials and testing tools to provide feedback for students. She also produced a nail-biting, fully online mid-semester test that was completed successfully as one of the first performed within the university, with students located across two small computer labs and the website providers standing by for emergency rescue. Having learned HTML, she programmed the department's old DOS-based tutorials into the Web format while updating the content and adding pictures and animations. As coordinator, she was heavily involved in the development of the new Melbourne Model curricula for second-year subjects, including a favoured elective subject called biochemical regulation of cell function. Stanley continues to deliver about half of the lectures in both subjects, which are very highly rated by her students.

As chair of the Biochemical Education Group of the Australian Society for Biochemistry and Molecular Biology for five years, Stanley organised workshops and initiated the inclusion of Biochemical Education Symposia in the annual COMBIO conferences. She has served on the Science Faculty's Academic Programs Committee since 2004, and since 2005 has been the health and safety representative for staff based on the Parkville campus. She is a member of the health and safety committees of the Faculty of Medicine, Dentistry and Health Sciences and the university. As well as counting herself a lover of the arts and an enthusiastic traveller, Irene Stanley is a keen follower of Formula One motor racing.

Graham Royston Parslow (1948–) took his BSc (Hons) and his Dip. Ed. from Flinders University in 1970. He took his MSc from the University of Adelaide in 1978 with a thesis on the purification of δ-amino laevulinic acid synthetase, while working from 1975 to 1977 as a high school teacher of science and biology and then lecturing part-time at the South Australian Institute of Technology in 1978. From 1978 to 1989 he held the position of senior tutor in the Department of Biochemistry at the University of Adelaide. During this time, he spent a year's study leave as a visiting fellow with E.J. Wood (1941–2008) in the Biochemistry Department of the University of Leeds. From 1989 to 1991 he was the coordinator of educational technology at Bond University School of Science and Technology. He came to the University of Melbourne in 1991 as a lecturer and has been an associate professor since 2009.

Parslow's appointment was specifically intended to improve the department's undergraduate teaching, especially to medical students, through the development of computer-assisted learning. This was an initiative of W.H. Sawyer, who had been involved in computer-based teaching since 1973. Generous funding came through the Mainstreaming the Digital Revolution program initiated by Alan Gilbert (1944–2010), Vice-Chancellor of the University of Melbourne from 1996 to 2004.

Graham Parslow won the inaugural Excellence in Teaching Medal of the Faculty of Medicine and the Invitrogen Teaching Award

The transformation of the department's offerings was validated in 1998 when Graham Parslow was presented with the inaugural Excellence in Teaching Medal of the Faculty of Medicine, and later, in 2009, with the Invitrogen Teaching Award. The latter is given annually by the Australian Society for Biochemistry and Molecular Biology and Parslow received it for excellence in teaching, extensive publication in the field of biochemical education and pioneering contributions to multimedia education. The award comes with a grant of $3000, which he used to visit Myanmar and contribute to the reconstruction of teaching facilities totally destroyed in 2008 by Cyclone Nagit.[4]

Graham Parslow's most significant publication consisted of three computer disks and accompanying documentation: *The 'Q' Instruction Package* was a pioneering contribution to the establishment of global e-learning.[5] Parslow is an associate editor of the journal *Biochemical and Molecular Biology Education*, with responsibility for features on multimedia and computer applications to education. He has been a regular contributor to the journal since 1984 and is the author of numerous conference papers and book chapters. He was editor of the newsletter of the Australian Biochemistry Society from 1993 to 1997.[6]

Leon Helfenbaum (1955–) came to the department, like so many others, with a background in agricultural science. Educated at Elwood High School, he spent two 'gap' years at the Institute for Youth Leaders from Abroad in Jerusalem before enrolling in science at Monash University, from which he took his BSc (Hons) with first-class honours in genetics in 1979. The next nine years were spent in Israel: first at the Ruppin Academy of the Israeli Ministry of Agriculture, where he completed courses in agricultural management and agricultural entomology, and subsequently at Kibbutz Yasur. In 1985 he was the technical and scientific adviser to the Field Crop Team, leading it when it was named the Outstanding Field Crop Team Western Galilee. Returning to Monash in 1989, he worked first as a research assistant in Phillip Nagley's laboratory and then as an administrative assistant in the Centre for Molecular Biology and Medicine and, in 1995, as a research associate. His work included investigation of the bioenergetic function of mammalian mitochondrial respiratory chain complexes, establishment of a biochemical assay to measure proton pumping, assessment of coenzyme Q analogues in proton pumping and membrane potential, and

Leon Helfenbaum at work in the Gething Laboratory, 2003

establishment of assay to assess the uncoupling activity of biologically active compounds.

Helfenbaum took his PhD from Monash in 1995 and came to the University of Melbourne the following year.[7] He joined Mary-Jane Gething's laboratory, working for the next eight years on protein structure, function, recombinant expression and engineering and signal transduction in eukaryotic cells. A paper on his investigations into the role of protein degradation on the regulation of the unfolded protein response, and how environmental stimuli modulate the response through activation of the SCF Cdc4 pathway, was published in 2007.[8]

In 2004 Leon Helfenbaum became a senior tutor, coordinating second-year biochemistry practical classes, and in 2009 he took over coordination of the third-year class.

Amber Willems-Jones (1976–) returned to the department where she had taken her BSc (Hons) in 1998, and her PhD entitled 'Investigation of the Molecular Interactions of the TyrR Regulatory Protein from *Escherichia coli*' in 2004. In 1999 she was named in the Dean's honour roll. From 2004 to 2011, Willems-Jones was a research officer at the Peter MacCallum Cancer Centre, working part-time on a familial cancer project with Joe Sambrook: they used computer models to assign risk of developing cancer for BRCA1 and BRCA2 mutation carriers from the kConFab study. In 2005 she began cancer research with kConFab – a national familial breast cancer consortium – in which her role was to genotype all family members with BRCA1, BRCA2 and p53 mutations, and 'look' for genomic rearrangements in these genes where no point mutations had been identified.

Amber Willems-Jones has worked with kConFab, a national familial breast cancer consortium

kConFab is an organisation of geneticists, clinicians, surgeons, genetic counsellors, psychosocial researchers, pathologists and epidemiologists from all over Australia and New Zealand, and which began enrolling families with a strong history of breast and breast/ovarian cancer in 1997. Genetic, epidemiological, medical and psychosocial data collected from these families by kConFab nurses and researchers are stored in a deidentified fashion in a central relational database. Biospecimens collected from family members are used to characterise germ-line mutations in predisposing genes such as BRCA1 and BRCA2. kConFab has accumulated data on more than 1400 multigenerational, multicase kindreds. While not a research organisation itself, kConFab aims to provide a resource for researchers by making data and biospecimens widely available to researchers for use in peer-reviewed, ethically approved, funded research projects on familial aspects of breast cancer. It supplies biological specimens and data to over 100 research projects worldwide.

Willems-Jones also carried out a small research project investigating a number of characteristics involved in survival and risk in prostate cancer patients within the kConFab cohort, which resulted in a number of publications.[9]

One member of the department's academic staff made an unusual move in physical terms by opting in 2013 to spend most of his time based in the Medical Building and the remainder with the research team with whom he had worked (in the Russell Grimwade Building and later Bio21) since 1999. Terrence Damian Mulhern (1969–) was born in Townsville and took his BSC (Hons) in Chemistry in 1990 and his PhD in 1995 from the University of Queensland. During that time he won the University of Queensland CSR Mills Group Prize in Chemistry in 1990, which is awarded to the final-year honours student considered to have attained the highest standard; an Australian Postgraduate Research Award in 1991; and the Student Poster Prize at the Lorne Conference on Protein Structure and Function in 1994.

Although Mulhern did not study biochemistry at university, his reading during his honours year inspired his move into a structural biology project with Susan Pond of the Department of Medicine of Queensland University in 1991. The aim of the project was to develop a recombinant antibody fragment as a therapy for poisoning by the herbicide paraquat (methyl viologen). After purifying and characterising the anti-paraquat antibodies and fragments (IgG, Fab and scFv), Mulhern was to use nuclear magnetic resonance (NMR) to study their interaction with paraquat. To do this he moved to Melbourne to work with Raymond Norton at the Biomolecular Research Institute in Parkville (where Peter Colman and Mark von Itzstein were developing the flu drug Relenza) from 1993 to 1995. Difficulties with protein expression hindered progress and the scope of the paraquat project, so Mulhern moved to a project using NMR to determine the structure of a malarial antigen (MSP-3) in collaboration with Robin Anders (then at the WEHI) and his future departmental colleague Geoffrey Howlett.

Mulhern spent two years in Oxford working with Iain Campbell, before spending the next two at the University of Adelaide collaborating with Grant Booker, their work being enabled by the newly installed NMR spectrometer. He came to the University of Melbourne in 1999 having been awarded the third Grimwade Research Fellowship in Biochemistry, which lasted until 2004. During this time he worked in collaboration with Heung-Chin Cheng on Src kinases, obtaining funding from the Cancer Council in 2005 and NHMRC in 2006. He won a Yamigawa-Yoshida Fellowship from the International Union against Cancer in 2003. This award is made 'to enable cancer investigators from any country to carry out bilateral research projects abroad which exploit complementary

Terrence Mulhern, winner of the third Grimwade Research Fellowship in Biochemistry, was very involved in the move to Bio21

materials or skills, including advanced training in experimental methods or special techniques'. Mulhern used it to train in the use of small-angle X-ray scattering with Professor Jill Trewhella at the Los Alamos National Laboratory in New Mexico, United States.

In 2004 he was offered a senior lecturer position in Biochemistry and some responsibilities that proved different from those he expected. As Mulhern recalled in 2013:

> The 2005 New Year dawned with a new baby, a new teaching job, a new building and lots of new responsibilities I hadn't really expected … I was the acting Scientific Director of the Bio21 NMR facility … I pretty much single-handedly dealt with the logistics of getting the facility up and running, liaising with builders, painters, plumbers, gas-fitters, training users, meeting OHS requirements and overseeing the installation of Dick Wettenhall's 'baby', the 800 MHz NMR, in the basement 'NMR Cave' below the Auditorium. It was more like a real cave than most people realise. Apparently, a builder had walked all over the 'waterproof' membrane prior to the tiling around the court-yard outside of the auditorium, so the NMR Cave roof leaked like a sieve when it rained and we actually had stalactites forming from the seepage through the concrete roof. The NMR machines were cov-ered in plastic tents and mosquitoes bred in the wheelie bins full of rainwater.[10]

Technical and managerial support for the NMR facility was eventually found and the roof was fixed, but not without a great deal of distress that would have been all too familiar to Trikojus.

NMR spectroscopy is a research technique that determines the physical and chemical properties of atoms or the molecules in which they are contained. It can provide detailed information about the struc-ture, dynamics, reaction state and chemical environment of molecules. The technique is used by chemists and biochemists to investigate the properties of organic molecules, although it may be applied to any kind of sample that contains nuclei possessing spin. Because of the range of information and the diversity of samples, including solutions and solids, NMR has made a considerable impact on the sciences.

Mulhern had been involved in teaching second-year students, but in 2013 he relinquished the role of subject coordinator in second-year bio-medicine and assumed that of coordinator of second-year biochemistry in the Bachelor of Science, while continuing his research in small-angle scattering.[11] His other research includes investigations of signal transduction and human diseases, linking protein structure, function and biology, and the study of protein structure and interactions by NMR.

Terry Mulhern's trajectory through the department is an unusual one, but of course the academic staff of Biochemistry and Molecular Biology who are based in Bio21 are not divorced either from their colleagues in the Medical Centre or from the students who are taught there.

The Grimwade Research Fellowship in Biochemistry requires special mention here. Four of the first six fellows spent much of their tenure in the Russell Grimwade Building, with the others coming to the depart-ment after the opening of Bio21. Funded from the bequests of Russell and Mab Grimwade, the fellowship provides funding for five years. The advertisement for the 2013 fellowship stated that it is:

> designed to provide the opportunity for an early career scientist to establish an independent research laboratory. The successful appli-cant will establish and lead an independent research program in the Department located within the Bio21 Institute, in any one of the following research areas: molecular cell biology, structural biology, protein biochemistry, molecular immunology and cellular imaging. Preference will be for applicants who seek to define the molecular basis of disease mechanisms or host-pathogen interactions. The successful applicant will be a recognized scholar as evidenced by high quality publications, capacity to attract research funding and presentations at conferences.[12]

The inaugural fellowship was awarded in 1992 to Phillip Dickson, a Melbourne Biochemistry graduate who went on to become Associate Professor and Deputy Head of the School of Biomedical Sciences and Pharmacy at the University of Newcastle. The focus of his work was the enzyme tyrosine hydroxylase, which controls the rate of synthesis of the catecholamines dopamine, noradrenaline and adrenaline. The title of his PhD, awarded in 1985, was 'Rat Transthyretin: Biosynthesis and Metabolism', and Dickson is one of the collaborators in the investigations of the Neurodegeneration Division of the Florey. During his time as a Grimwade research fellow, his work concentrated on the engineering of recombinant proteins, an area of great importance in the field of protein pharmaceuticals, as they can be produced in large amounts and rapidly and easily purified.

Phillip Dickson, inaugural winner of the Grimwade Research Fellowship in Biochemistry in 1992

The second Grimwade research fellow was something of an anomaly, being far from an early-career researcher and appointed for an interim period only. He brought with him, moreover, a highly significant piece of equipment – a 500MHz high field NMR spectrometer. Professor Robin Bendall moved to the department following the reorganisation of the Department of Chemistry and Biochemistry at James Cook University, where he had been Nevitt Professor of Chemistry and Head of Department since 1985. The results of Bendall's work on the development of NMR methodology during his tenure of the Grimwade fellowship – he spent a good part of his two-year appointment in the United States – were published in 1998.[13]

Terry Mulhern was the third Grimwade research fellow, and following him in 2004 was Anthony W. Purcell (1967–), who had taken his BSc (Hons) in 1987 and his PhD in 1993 from Monash before moving to Adelaide and Flinders universities for postdoctoral studies. He subsequently moved to the Department of Microbiology and Immunology at Melbourne University in 1997, where he continued work as a postdoctoral fellow with James McCluskey on immune recognition and the key role peptides play in modulating health and disease.[14] He was awarded a C.R. Roper Fellowship in 2003 and began to apply cutting-edge mass spectrometry techniques to key questions in immunology and, of course, biochemistry in general. In 2005, having been awarded the Grimwade

Research Fellowship, he moved to the Department of Biochemistry and Molecular Biology in Bio21. He was awarded an NHMRC Senior Research Fellowship in 2008 and was appointed reader in 2009. Tony Purcell returned to Monash as a professorial fellow and Head of Quantitative Proteomics under the Talent Enhancement Scheme in 2012.

During his time with Bio21, Purcell investigated the biomolecular structure of cells crucial to the functioning of the immune system. Information gathered from this

Anthony Purcell, winner of the 2005 Grimwade Research Fellowship in Biochemistry, became Head of Quantitative Proteomics at Monash University in 2012

investigation will guide the selection of components for vaccines against chronic viral illnesses (such as HIV and hepatitis) and cancers. His research team was successful in attracting over $12 million in research funding, including eleven NHMRC project grants, two ARC linkage grants, three ARC LIEF grants and one ARC discovery grant. The team was a major collaborating partner in larger projects which received funding of over US$2 million from the US National Institutes of Health and two international grants from the Juvenile Diabetes Research Foundation to research type I diabetes. Once again, the state-of-the-art mass spectrometry facilities at Bio21 were crucial regarding the capacity of the university to play an important role in this project. One of Purcell's most-cited papers of this period defined a new mechanism for adverse reactions to pharmaceuticals that has represented a paradigm shift in drug hypersensitivity research.[15]

Unlike Tony Purcell, Danny Martin Hatters (1974–) took both his 1996 BSc (Hons) and his 2002 PhD on characterisation of human apolipoprotein C-II amyloid fibrils from the University of Melbourne. He undertook postdoctoral work with Karl Weisgraber at the Gladstone Institutes at the University of California, San Francisco, investigating how three variants of apolipoprotein E – apoE2, apoE3 and apoE4 – differ in their conformation and biophysical properties as a basis for understanding the mechanisms underlying the elevated risk that the apoE4 isoform confers for Alzheimer's disease. He had already won the 2001 Young Biophysicist Award of the Australian Society for Biophysics, given to an Australian biophysicist with no more than five years' postdoctoral experience. In 2006, having been promoted from postdoctoral fellow to research scientist at the Gladstone Institutes, Hatters won both the Boomerang Award and the Young Investigator Award.[16] Returning to Melbourne in 2007 with a C.R. Roper Fellowship, he was awarded the fifth Grimwade Research Fellowship in 2009 and took out the annual Applied Biosystems

Danny Hatters won a Grimwade Research Fellowship in Biochemistry in 2009 and awarded an ARC Future Fellowship in 2012

Edman Award of the American Society for Biochemistry and Molecular Biology (ASBMB). In 2012 he was appointed to the teaching and research staff of the department and was awarded an ARC Future Fellowship.

The major area of research of the Hatters laboratory is Huntington's disease, with other programs in motor neuron disease and cancer. The laboratory explores how proteins change conformation in cells as part of normal and pathogenic processes, including those associated with misfolding and aggregation. Key elements of this research include the development and application of new, fluorescence-based probes and biosensors to view protein dynamics in cells.[17]

Gregory William Moseley (1974–) was awarded the Grimwade Research Fellowship in 2013. His PhD was awarded by the University of Sheffield in 2002, based on research there and the WEHI investigating the structure and function of tetraspanin proteins, in particular their roles in immunity and cancer. Having been awarded a Royal Society Fellowship in 2001, he joined the Austin Research Institute in Melbourne and subsequently moved to the Department of Biochemistry and Molecular Biology at Monash, where he established an independent

research program investigating the immune evasion strategies and pathogenic mechanisms of viruses, notably lyssaviruses such as rabies and
Australian bat lyssavirus, as well as paramyxoviruses including Nipah
and Hendra viruses.

Gregory Moseley seen here
with Angela Harrison, won
the Grimwade Research
Fellowship in Biochemistry
in 2013 and worked on the
immune evasion strategies
and pathogenic mechanisms
of viruses

A significant social occasion that has brought the staff in the Bio21
Institute and the Medical Building together for over two decades is the
Third Year Staff–Student Dinner which takes place in second semester
every year. This was initiated by Bruce Grant. From the 1980s to the early
1990s, the practical laboratory teaching and preparation staff organised
end-of-semester barbecues for final year biochemistry students in the
garden of the Russell Grimwade Building. These were well attended by
both staff and students, but the celebrations sometimes became rather
boisterous, perhaps because of the freely available wine and beer. Grant
thought it might be preferable for academic staff and final-year students
to meet in a setting that, while still encouraging good communication
between academics and students, was slightly more formal. He therefore
organised career nights at which scientists from within and outside the
university talked about their work and how they came to be involved in it.

From these came the first of more than twenty staff–student dinners held at University House, at which final-year biochemistry students enjoy a three-course meal with the academic staff and practical class demonstrators and have the opportunity to hear an invited speaker. The speakers were initially invited from outside the department, some of them former members of staff: among the early speakers were the noted scientists Edwina Cornish and Geoffrey Tregear. More recently, speakers have included younger researchers and postgraduate students from the department. Between a third and half of the eligible students attend. While the occasion is essentially a social one, the custom of having staff move from table to table for successive courses ensures that diners have the opportunity of discussing life as a biochemist and other matters with most of the academics present. In 2013 the speakers were Diana Stojanovski, who heads Biochemistry's Mitochondrial Biogenesis and Disease laboratory, and Silvia Teguh, a graduate student in the Tilley laboratory who is researching new antimalarial drugs.

Bruce Raymond Grant (1936–) took his BSc (Ag) from the University of Queensland in 1958 and both his MSc in 1960 and PhD in 1962 from Purdue University, Indiana. From 1962 to 1964 he was a research fellow in the Department of Cell Biology at the University of California, Berkeley, returning to Australia in 1964 and taking a position as research scientist (later senior research scientist) in the CSIRO Division of Fisheries and Oceanography. He spent 1968 to 1972 as visiting fellow and Assistant Professor in the Biology Department of Queens University, Kingston, Ontario and received a joint appointment in the School of Agriculture and the Department of Biochemistry and Molecular Biology in 1972 as a senior lecturer. He was promoted to reader in 1985 and retired in 1997.

Grant, who had been awarded the Australian Institute of Agriculture Research Prize and the Fulbright Travelling Fellowship in 1958, held a Purdue Graduate Fellowship from 1958 to 1962. In 1982 he was visiting fellow at the Australian Institute of Marine Science in Townsville. He spent 1986 in the Department of Plant Sciences at Oxford, 1987 as visiting professor at Drexel University, Philadelphia, and 1994 as NSERC visiting fellow back at Queens University. Much of his work was done in collaboration with colleagues in the School of Botany, including his most-cited paper on the complex action of phosphonates as antifungal agents.[18]

The paper discusses two hypotheses: the first, a straightforward fungitoxic mode of action, similar to that found in other systemic fungicides; the second, a novel and complex mode of action which includes components of direct anti-fungal activity and a stimulated host defence response from the infected plant. It concludes that phosphonates stop pathogens growing in plants by slowing the growth of the pathogen and inhibiting its sporulation through direct fungistatic action. This reduction in pathogen growth rate allows the host defence system extra time to develop and to kill the

Bruce Grant inaugurated the annual staff–student dinner and worked with, among others, Gretna Weste of the School of Botany

invading organism. Phosphonates alter the metabolism of the pathogen in such a way that a more rapid, more effective defence response can be mounted by the host. The paper notes that this complex mode of action accounts for the variation in results using different plants and pathogen strains or species, and suggests the reason why resistance to phosphonates has not been found in practice. Resistance to fungicides is associated with a mutation at a single site in the pathogen, because fungicides usually inhibit a single enzyme or process in the pathogen. It appears likely that phosphonate may have multiple biochemical sites of action.

An earlier project, with Gretna Weste of the School of Botany into *Phytophthera cinnamomi* (cinnamon fungus), an organism that can kill orchard and garden plants as well as native plants such as eucalypts, was supported by the William Buckland Foundation. It investigated the factors that dictate whether the fungus will destroy forests or coexist with them. The research resulted in a number of co-authored papers.[19] A colleague, writing in 1985 that Grant had developed expertise and recognition in two fields, algal biochemistry and physiology, as well as the biochemistry and physiology of fungi and fungus–plant interactions, commented that his work on the role of calcium and carbohydrate in

Phytophthora zoospore development and differentiation might prove to be a major 'bombshell' in the field.[20]

When Michael Austin Tuohy (1954–), who took his BSc (Hons) in Botany from Monash in 1978, was appointed as a technical assistant in the School of Botany in 1980, he could have had little idea of how his role and responsibilities would change over the next twenty-nine years. In 1982 he was promoted to technical officer grade 2 in charge of the large plant cell biology research laboratory. The requirements of the position included, in addition to an honours degree in botany combined with a good background in chemistry,

Michael Tuohy, the business manager who revolutionised the department's storekeeping, was highly involved in the move to Bio21

biochemistry or microbiology, extensive experience in microscopy, photography and sectioning of plant-embedded tissue. A driver's licence was essential and a scuba-diving licence considered an advantage. In 1981 he had already been identified as the co-author of a paper with Bruce Knox on pollen aerobiology in southern Australia, presented at the Australian Weeds Conference.[21]

Tuohy remained with the Botany School until 1990, when he was appointed laboratory manager grade 4 in Biochemistry. In 2000 he was named departmental manager. Michael Tuohy was awarded a Melbourne University Bronze Medal in 2005. The citation by Kenwyn Gayler read in part:

> His performance as Business Manager has been outstanding during a difficult period of transition for the Department into a complex new infrastructure. He has made a major contribution to the restructuring of the operation of the department into a two-site operation, the Bio21 Institute site and the Grimwade site, and his attention to detail in the complexities of the new budget has been invaluable.[22]

Well before the opening of the Bio21 Institute, Tuohy had implemented a reorganisation that was to save the department several thousand dollars a year. Trikojus had noted in 1967 that Graeme Strange had revolutionised its storekeeping. By the time Tuohy arrived, the system he had put in place was resulting in a considerable loss each year. All staff were required to sign out for all supplies, including pencils and pens; the store was open only for limited times; and orders from outside suppliers needed to be presented to Strange, sent to another floor to be typed, signed by the laboratory head, presented to the Head of Department, approved and returned to Strange before they could be sent out. Tuohy streamlined all the processes and established a liaison between other departments that enabled exchanges of chemical supplies, among other things. In a couple of years, the store was generating income for the department and saving both academic and professional staff a good deal of time.

The transition to Bio21 took a toll on Terry Mulhern, and it had a similarly severe impact on Michael Tuohy, who noted that:

> Bio21 is not simply the culmination of past work. It now involves a complete revamping of effort to make it work. The coordination to move 100 of the research staff and students, and their scientific equipment, into the new facility has been significant. There have been substantial changes to policy and procedures to be able to operate in two locations. The greatest workload has been to try to function effectively within the Bio21 guidelines (as per the Penington report) with limited technical and administrative staff. The move, the restructuring and the changing policies and guidelines have all been in addition to my normal workload. This year so far has been the most difficult in 25 years in the university both physically, mentally and emotionally.[23]

The departmental manager's responsibilities were divided into six principal areas, including assistance in policy and planning, which in part involved membership of a number of university and faculty committees concerned with everything from waste disposal to information technology; personnel management, notably in communication between the academic and non-academic staff and coordination of in-house and external training; financial and resources matters, including the preparation of departmental budgets covering several million dollars a year and planning for changes in laboratories, lecture theatres, computer networks, NMR suites and laboratories for computer-assisted instruction; marketing and promotion, including the coordination of symposia, Discovery Day activities and so on; responsibility for environmental health and safety procedures and compliance, and, while keeping an eye out for possible sources of research funding, assessing the legal and financial implications of any new venture. Finally, Tuohy's duties specifically included assisting the director of the Molecular Science and Biotechnology Institute of Bio21 on matters relating to the planning of the institute.

From 1993 to 2009, Tuohy was assisted by Martin Greasley (1956–), initially as chief technical officer (Electronics and Computer Workshop) and, from 2005, as manager of information technology. He had a lot to do. Greasley, who took his BSc (Microprocessor Applications) in 1905 from Deakin University, had also taken several other qualifications, including an Associate Diploma of Engineering from Monash University, a Certificate of Technology from Moorabbin College and a Communications and Electronics Certificate from RMIT. He had previously worked at Lyndhurst Radio Station and in various parts of Telecom. The

Martin Greasley was manager of information technology at a time of rapid change in demand

duties of his new position included responsibility for the management and general operation of the department's Electronics and Computer Workshop and centralised personal computer facilities, provision of expert technical support on all electrical equipment, regular inspection and maintenance of all electrical teaching and research equipment, and both suggesting and undertaking appropriate design modifications.

When the department was split between the teaching staff operating from the Russell Grimwade Building and later the Medical Building, and those who moved to Bio21, Greasley's position was reclassified to take account of the additional responsibilities he incurred and to recognise both the university-wide impact of the position (because of the need to align department and institute requirements with those of the whole institution) and the necessary management and technical issues. The specific changes in responsibility were defined by the Head of Department as:

- Responsibility for promulgating strategic directions in IT services to both the Department of Biochemistry and Molecular Biology and the Bio21 Molecular Science and Biotechnology Institute.
- Management of IT services to staff and students in both the above locations.
- Increasing responsibilities for the greater complexities of new and emerging technologies.
- Interpretation of policies, planning for the academic year and managing periodic reviews of IT in the two locations.[24]

Since leaving the department at the end of 2009, Martin Greasley has been able to enjoy his interest in sailing, notably with the Queenscliff Cruising Yacht Club.

The multifaceted nature of the career of Tim Anning in Biochemistry is encapsulated in a recollection from his time as senior technical officer/manager of the animal facility. In 1995 Timothy Lyle Anning (1959–) was elected to the University Council. At the time, he was in his penultimate year in Biochemistry, coming to the end of a period in which he had been the only staff member employed in the animal facility. He remembered scrubbing out animal cages until shortly before 4 p.m., changing into a suit and walking across to the august surroundings of the Council Chamber in the Old Quadrangle to take his seat at its deliberations. He was still a member when Robert Hannaford was commissioned to paint the group portrait of the 2003 council, in commemoration of the university's first 150 years: he can be seen seated third from the right at the council table.[25]

Anning, who took his BSc in 1987, had joined the department in 1978. He worked as a technical assistant in the undergraduate teaching laboratories until the end of 1984, moving to the animal facility where, in November 1989, he took over from Ian Young. The animal house was on the fourth floor of the Russell Grimwade Building, but then postgraduate activity transferred to the David Penington Building, and the Biological Research Facility under the management of Max Walker, supporting work in several medical departments as well as CSL Limited and other industry partners, moved in 2005. From 1989 to 1997, Anning was in sole charge of the animal house, a situation that required his presence there for part of every day, including weekends and public holidays, since the animals required daily attention. As manager of the facility, he was responsible for the entrance

Timothy Anning was simultaneously manager of the animal house and a member of the University Council and executive officer (animal ethics and biohazards)

and exit of all animals and their welfare. (Many members of staff contribute to a minor aspect of the latter. The cardboard tubes inside toilet rolls are the perfect size for small animals to run through or hide in, and new, autoclaved ones are always needed, so a collection point exists in the Medical Centre as well as Bio21.)

Anning's personal concern for animal welfare was manifested in his undertaking the position of executive officer (animal ethics and biohazards) on a part-time basis from June 1996 to June 1997, spending half his time in the Russell Grimwade School and half in charge of animal ethics in the Research Office. In July 1997, he became executive officer, animal ethics. In 2007, when he had served for twelve years on the University Council, the maximum term allowable, Tim Anning was presented with a Silver Medal. The citation noted that:

Particular mention should be made of his long service to the Legislation and Trusts Committee and its predecessors, where not only his eye for detail but his awareness of the general context of University policy behind the legislative instruments have made his contributions invaluable. Members will recall the not infrequent occasions when court commitments have prevented the Chairs of those committees from attending Council to present the legislation, when Tim has stepped in to do so with great clarity and competence. His service on other committees of Council, notably the Building and Estates Committee and the Council Nominations and Governance Committee, has also been greatly appreciated.[26]

As the citation also noted, animal ethics and gene technology are areas of high public interest, both to government agencies and to the animal welfare and other lobbies. In this context, it is worth noting the university's commitments and policy. Responsibility for managing ethics and related approval processes, and for developing and implementing policies and guidelines for responsible research, rests with the Office for Research Ethics and Integrity. Animal ethics approval is required for all university teaching and research activities that involve the use of animals. This covers the whole spectrum of research and teaching activity, from laboratory-based study to the observation and noting (without handling) of animal behaviour in its own habitat. A new Australian code for the care and use of animals for scientific purposes, to which all users of animals in research and teaching are required to adhere, was released by the NHMRC on 23 July 2013.[27]

When Jacqueline Munro-Smith began work in the Department of Biochemistry and Molecular Biology in 2000, she was no stranger to the university. Munro-Smith had worked in various positions since taking her BA from Monash University in 1981, principally in the Department of Human Resources, where she held positions of increasing responsibility until 1992, when she joined the Victorian College of the Arts as arts administrator, secretary to the council and affirmative action officer, before moving back to Human Resources in 1999. Between 1995 and 1998 she decided to broaden her experience through contract work, which involved long-term temporary appointments with Faulding Healthcare, Linfox, CSL Bioplasma and VENCorp. At the latter she won an award for service during the emergency caused by the explosion at Longford, a catastrophic industrial accident which occurred on 25 September 1998, killing two workers. Gas supplies to the state of Victoria were severely affected for two weeks and some staff worked round the clock for several days. Among the many highlights of her temporary work, Christmas lunch in the Linfox boardroom, served by Lindsay Fox himself, is a special memory. Jacquie Munro-Smith studied her Graduate Diploma in Women's Studies at Victoria College, Rusden in 1988. She is currently studying mental health theory online, with a view to undertaking volunteer work with the Samaritans or a similar organisation when she retires to Cornwall to 'run a coffee shop and in-between customers, write The Great British Novel'.

Jacqueline Munro-Smith, administration manager and personal assistant to the Head of Department, is also well known for her emails reminding casual staff to submit time sheets

Jacquie Munro-Smith's emails reminding casual staff of the need to submit time sheets were highly imaginative and eagerly anticipated. The following example, sent in 2013, is one of dozens:

I met a man who had a PhD
Do you know what he said to me?
'I wouldn't be where I am today
If it hadn't been for my casual pay.'
I met an RA on the stair
Healthy glow, shiny hair
'It's thanks to pracs', she said with a grin.
'I keep putting my timesheets in.'
I met happy folk all over the place
Wealthy, with big smiles on their face
But why? What's the premise?
Putting your casual hours in Themis.[28]

The duties of the administration manager and personal assistant to the Head of Department are varied, encompassing responsibility for providing high-level secretarial and administrative support, acting as minute secretary to the departmental committee, liaising with staff of the department and other sections of the university in relation to departmental and university procedures and operations, graduate administration (including admissions), managing departmental events and the work-experience program, looking after visiting fellows and students, and acting as publications coordinator.

One other person experienced the transformation of the Department of Biochemistry and Molecular Biology and the development of Bio21 in a very particular way. Richard Edward Hugh Wettenhall (1943–) took his BSc from Melbourne in 1966 and his PhD, on the influence of insulin on protein synthesis in connective tissue, from Monash in 1970, before spending the next four years in the United States and England on postdoctoral fellowships at the universities of Chicago and Birmingham. From 1973 to 1976 he worked at the Australian National University as a research fellow in the Molecular Biology Unit of the Research School, and from 1976 to 1988 as senior lecturer and later reader in the Department of Biochemistry at La Trobe. From 1986 to 1988 he worked in the United States at Genetech USA, the California Institute of Technology and the University of Texas, Austin. In California he worked with the pioneer in systems biology and one of the initiators of the Human Genome Project, Leroy Hood (1938–), whose research laid the foundations for studies in proteomics and genomics. This work resulted in a number of papers on sequence analysis.[29]

Wettenhall took up the position of Professor of Biochemistry and Microbiology at Melbourne in 1988. During his time as Head of Department, his research focused on medical biotechnology, notably on the structure and function of proteins and RNAs. His most-cited paper from this phase of his research career reported investigations into the involvement of a gonadal protein termed inhibin

Richard Wettenhall (*right*) with Bill Sawyer and Gerhard Schreiber in 2001, was the inaugural director of the Bio21 Institute

in the control of FSH secretion. Using a combination of techniques (gel filtration, reversed phase HPLC and SDS-PAGE), a highly potent, biologically active fraction was isolated which was capable of suppressing FSH cell content in an in vitro bioassay.[30]

During the 1990s, Wettenhall was increasingly involved with the establishment of the Bio21 Institute. The $400 million project was announced by the Victorian Premier in 2000, with Wettenhall as interim director and the university's Vice-Principal (Property and Buildings), Doug Daines, as capital projects manager.[31] In his role as the inaugural director, Wettenhall led the development of multidisciplinary research programs and related platform technologies, including metabolomics, mass-spectrometry-based proteomics, NMR spectroscopy, analytical and synthetic chemistry and electron microscopy.

Like Bill Sawyer, Dick Wettenhall, whose family runs a Victorian Hereford cattle and sheep farm, established a winery. The Gurdies Winery in South Gippsland has won awards for its produce, which include both table and dessert wines. Dick Wettenhall retired from the university in 2007, having overseen the transfer of the department's research activity to Bio21. In 2013, the Bio21 Institute announced the Wettenhall Establishment Award. This annual postdoctoral award, designed to assist the recipient's transition from senior postdoctoral scientist to independent researcher, supports a person with a strong research track record who is currently undertaking work aligned with one or more of the research themes of the Bio21 Institute. The inaugural recipient of the award was Darren Creek, whose research focuses on drug discovery for tropical diseases caused by the parasitic protozoa *Plasmodium falciparum* and *Trypanosoma brucei*. The goal of his major project to expand our understanding of parasite metabolism and the impact of drugs on metabolism was to facilitate new drug discovery and optimise the utilisation medicine based on a mechanistic understanding of drug action.

The opening of the Bio21 Institute took many people from the southern part of the university campus to the western end. There were many scientists who were already in the department when the move took place, and whose work deserves some attention.

Four Heads of the Department of Biochemistry – *left to right:* Mary-Jane Gething, Lloyd Finch, William Sawyer and Kenwyn Gayler

CHAPTER 10

The New World

I N 2005 THE Department of Biochemistry and Molecular Biology was split when the research staff and students moved from the building which had housed undergraduate teaching since 1958, and the whole department since 1961, to the newly opened Bio21 Molecular Science and Biotechnology Institute in the David Penington Building.

The Bio21 Institute is located in the university's Western Precinct. Bounded by Flemington Road, Park Street and Story Street, the site would have been familiar to Biochemistry staff and students as the location of the School of Veterinary Science, which had been there since 1908. Before that date, this area had accommodated a livestock market, a fact recalled in the names of several hotels, such as the Horsemarket, that stood along Royal Parade until well into the twentieth century. In 1930, CSIRO established the Animal Health Research Laboratory on the site. That moved to Geelong in 1996 and one of the former CSIRO buildings now serves as the Bio21 Institute Business Incubator. The Bio21 Molecular Science and Biotechnology Institute building is a large one: the *Annual Report* for 2003 noted that it added 23 000 square metres to the university's property portfolio.[1]

The move of part of the department to the Western Precinct involved a greater geographical division than when Trikojus complained of having parts of his department accommodated 400 metres from each other. Older staff, however, would have remembered decades when various schools and departments of the university had operated across the not inconsiderable barrier of Royal Parade. Biochemists, moreover, had a long history of close collaboration with staff and students of the School of Veterinary Science. Although the loss of a common tearoom and daily contact in corridors, stairwells and offices significantly changed the dynamic of the Department of Biochemistry and Molecular Biology, staff in Bio21 are able to maintain regular contact with undergraduate students through their work as coordinators of over the dozen subjects offered.

A little background on Bio21 is appropriate here. The university announced in its *Annual Report* for 2000 that it:

> determined to use the proceeds of the Melbourne IT 'float' to pursue important strategic goals. First, it set aside $45 million to invest in a new Institute of Molecular Science and Biotechnology ... The Institute will operate within Bio21, a partnership linking the State Government, Walter and Eliza Hall Institute of Medical Research, the Royal Melbourne Hospital and the University in a major commitment to bio-molecular science and biotechnology. Bio21 will be dedicated not just to basic and strategic research in biomolecular science, but also to facilitating a concentration of infrastructure, expertise and venture capital sufficient to encourage the development, clinical trialling and commercialisation of key research findings in biotechnology. The aim is to make the Bio21 precinct one of the most significant 'critical mass' concentrations of biotechnology development in the world.[2]

Early planning documents identified many functions for the institute, from multidisciplinary research involving the university, Bio21 partners and industry to fostering and mentoring start-up enterprises, providing industry-targeted education and training, and the commercialisation of intellectual property generated by the institute.

The design of the David Penington Building itself (it was devised by DesignInc and constructed by Baulderstone Hornibrook) is inspired by the cell nucleus and the associated structure is interpreted figuratively throughout the complex in the form of 'breakout spaces'. It won a slew of awards, notably the Science Industry of Australia 'Lab of the Year', the RAIA Marion Mahony Interior Architecture Award, the Award of Commendation for Interior Spaces of the Illuminating Engineers Society for 2005, and both the Public Building Award and Overall Award of the Property Council of Australia (Victoria) for 2006.

Key to the 'critical mass' mentioned earlier was the establishment within the institute of a number of platform technologies concentrated in a few major areas of strategic importance beyond the financial capacity of individual research groups or disciplines. Each facility is accessible to outside academic and industry research groups. The NMR facility incorporates nine instruments, some previously owned by the university or the WEHI, and others, like the $5 million, 800-megahertz NMR itself, more recently acquired. Mass spectrometry, which has emerged as a core analytical chemistry platform for proteomics and metabolomics, is enabled by twelve mass spectrometers. The Electron Microscopy Unit and clean room constitute a key facility designed for applications in the physical sciences, life sciences and engineering in the increasingly important area of nanotechnology. The unit is equipped with five state-of-the-art electron microscopes, available to academic and industry users on a fee-for-use basis. Eight research groups of the institute are concerned with synthetic chemistry, with applications such as making molecules, including candidate drugs. The institute accommodates a rodent house that includes both clean-room breeding facilities and general experimental working areas and a Drosophila (fruit fly) breeding facility which provides a resource for investigating functions of counterparts of human cell signalling components. Bioinformatics is an area of increasing interest.

The research themes of the institute are listed on its website as:

structural and cell biology, which provides an understanding of the organisation of complex biological systems and molecular processes that underpin normal cellular development and disease; chemical biology, the small molecules that impact on biological systems and environmental ecosystems or can be used to manipulate biological processes to provide the basis of novel therapeutics and insect control agents; and nanobiotechnology that brings together the physical and

life sciences with engineering, working at the sub-nano scale, to provide a new level of health, agricultural and environmental research.[3]

The staff of the Department of Biochemistry and Molecular Biology
undertake research activities in one or more of six designated research
themes which have been actively strengthened in the department over
the past ten to fifteen years: molecular cell biology, protein-folding
diseases, host–parasite interactions, cell signalling, genomics and
molecular immunology.

The process of moving offices and laboratories from the Russell
Grimwade Building to the Bio21 Institute was a difficult one, as shown
by the stress it placed on both Michael Tuohy and Terry Mulhern. The
later move of the teaching staff and associated facilities from the Russell
Grimwade Building to the Medical Centre was also complex and further
complicated by being undertaken during semester, while classes were still
in progress. The same firm was involved in each move. Kent Moving &
Storage, established in 1946, is Australia's largest privately owned storage and relocation company. In the move from the Russell Grimwade
Building to Bio21, the design of the laboratories had been planned to
ensure that the correct three-phase, single-phase, 10-amp or 15-amp
power points, gas and other facilities were available in the right areas
and provided additional capacity for future needs. Michael Tuohy noted
that the freezers in the Russell Grimwade Building were carefully packed
and left running until the last possible minute before being loaded onto
a truck, transported and reconnected in Bio21 within two hours. Most
laboratories were able to resume operation within two days. Biological
safety cabinets were decontaminated and sealed a day or two before the
move to make them safe for transport, then moved, tested and recommis-
sioned within a few days.

Both the move to Bio21 and the later move of the undergraduate
teaching to the Medical Centre took place under the general direction of
Ken Gayler, who succeeded Mary-Jane Gething as Head of Department
in 2005. Kenwyn Ronald Gayler (1940–) took his BAgSci (Hons) from
the University of Adelaide in 1962 and his PhD from the University of
Queensland in 1969. From 1964 to 1970 he worked as a research scientist at the David North Plant Research Centre, CSR Ltd. He came to a

lectureship at the University of Melbourne in 1975, retiring in 2006 as associate professor. In 1980 he won a British Council Special Studies Program Award to work in the Rothamsted Experimental Station in England on the problem of poor nutritional quality of grain and seed crops. He successfully used the techniques that made it possible to study synthesis, processing and secretion of plant proteins in cell-free systems. In 1987 Gayler was IDP visiting adviser at the Prince of Songkla University in southern Thailand. IDP, which has undergone various name changes since its foundation, is an international-development program involving a number of Australian universities and colleges which supports over thirty institutions in countries such as Indonesia, Malaysia, Singapore, Thailand, the Philippines, Brunei, Papua New Guinea and the South Pacific.

Gayler had established his reputation well before he published the research which attracted the attention of the general public. His most-cited paper, on the effects of nutritional stress on the storage proteins of soybeans, was published in 1985.[4] It examined the effects of sulphur deficiency on the complement of proteins laid down in developing seeds of soybean (Glycine max L. Merr), finding that sulphur deficiency caused

Kenwyn Gayler and Bruce Livett at the demolition of the Russell Grimwade Building

a 40 per cent decrease in the level of glycinins and a contrasting elevation in the level of β-conglycinins. The subunit composition of these proteins was also affected, notably showing a three-fold increase in the β-subunit of β-conglycinins in the sulphur-deficient seeds that accumulated largely as the B0-isomer of β-conglycinins, a protein which while virtually devoid of methionine and cysteine retains the physical properties of a normal 7S storage protein. The paper demonstrated that a high degree of selectivity can be exerted by environmental stress over the accumulation of proteins in developing seeds.

It was, however, almost twenty years later that different investigations by Gayler and Bruce Livett were published and became widely cited in mainstream publications as well as in the professional literature. Livett and Gayler's work on the compound ACV1, extracted from a species of cone shell found on the Great Barrier Reef, was first reported at the second Venoms to Drugs conference held on Heron Island in 2002. These conferences, to date held in 1998, 2002, 2005 and 2011, have provided an international forum for the presentation and discussion of high-level research leading to the development of pharmacological tools and novel drugs from the venoms of cone snails, scorpions, spiders, snakes and other species. Gayler and Livett's findings suggested a medical application for ACV1 in the treatment of chronic pain.

ACV1 is a 16 amino acid peptide, a novel alpha-conotoxin identified by gene sequencing, and an antagonist of neuronal nicotinic acetyl-choline receptors that is active in suppressing the vascular response to selective stimulation of sensory nerves in vivo. Its initial use, reported as recently as 2013, is expected to be in the treatment of sciatica, shingles and diabetic neuropathy.[5] Reporting their work in *UniNews* in 2002, Livett noted that ACV1 treats pain by blocking its transmission in the peripheral nervous system, and that it acts on a different class of pain receptors to other conotoxins trialled earlier. It has the advantage of being able to be injected into the patient's muscle or fat layer rather than the spine.[6] One of Gayler and Livett's most-cited papers reports that a novel α-conotoxin identified by gene sequencing is active in suppressing the vascular response to selective stimulation of sensory nerves in vivo.[7]

Bruce Grayson Livett (1943–) took his BSc (Hons) and PhD from Monash in 1965 and 1968. He spent the following three years in England as a Nuffield Dominions demonstrator, working with Derek Hope on neurophysins, using immunohistochemistry to locate them in the hypothalamo-pituitary axis, and with James Parry on scrapie in sheep, discovering a new neurosecretory pathway from the hypothalamus to the median eminence that was rich in neurophysins. Returning to Monash, where he worked from 1972 to 1975, Livett developed an interest in chromaffin cells as models of neuronal function and created a method for isolating single cells from bovine adrenal medullae suitable for primary cell culture. In England he had also taken up bassoon playing, later teaching it in Melbourne. He returned to playing the piano in the 1970s while in Canada, where he was on staff at Montreal General Hospital Research Institute and McGill University, attaining the rank of full professor. He joined the group of Albert Aguayo (1934–) to carry out research on muscular dystrophy and on neuropeptides as neuromodulators.

Bruce Livett joined the Department of Biochemistry and Molecular Biology at Melbourne University in 1983 and built a research team that studied the actions of the neuropeptides substance P and the encephalins as neuromodulators. Together with Bill Sawyer, Graham Parslow and Peter Harris of the Department of Physiology, Livett spearheaded the introduction of computer-aided instruction into the science and new medicine course teaching programs. Later, he commercialised his research team's work on ACV1 by securing a licensing agreement with Metabolic Pharmaceuticals, Melbourne who took the invention through

Bruce Livett with cone shells

to phase IIA clinical trials for the treatment of painful sciatica. He was president of The Royal Society of Victoria from 2006 to 2007. Livett maintains his research website on cone shell and conotoxins and is an associate member of the department.[8]

In recent years, in what may be seen as a serious version of the tree change, Livett has been the owner and director of the Zebra Rock Gallery in Kununurra, Western Australia. Located on the Ord River, upstream from the Diversion Dam on the Victoria Highway, the gallery is a welcoming oasis for travellers and employs specialist carvers to create decorative items from rare ancient Kimberley siltstones: Zebra Rock, Primordial Rock and Liesegang Weather Rock (600, 1200 and 200 million years old, respectively). The property also has 400 mango trees, a workshop, a coffee shop, talking birds, a children's playground and fish feeding at the jetty. Zebra Rock Gallery won the 2012 East Kimberley Small Business Award.

It is time to turn to the work of some of the laboratory heads working in the department in 2013 who were appointed before 2012, beginning with a man who joined the department in the late 1970s. The other staff will then be introduced in the order in which they were recruited to the department.

Geoffrey John Howlett (1947–) took his BSc (Hons) in 1969 and his PhD in 1972 from Melbourne, leaving the following year for a twelve-month stint at the John Curtin School of Medical Research at the ANU, before departing in 1973 for a four-year postdoctoral fellowship in the Department of Molecular Biology and Virus Laboratory of Howard K. Schachman at the University of California, Berkeley. In 1976 he took a position as research fellow in the Department of Biochemistry at La Trobe University and in 1979 he joined the department at Melbourne as a lecturer. Promoted to senior

Geoffrey Howlett's research covers diseases such as dementia, diabetes and heart disease

lecturer in 1986, Howlett became associate professor and reader in 1993. He has spent frequent short periods of time in overseas laboratories in the United States and Great Britain. While at Berkeley, Howlett participated in two applications of the airfuge which was then being developed at Beckman Instruments, Inc. His interest in the airfuge was maintained after his return to Australia and he used his imagination, background and experimental skills in applying the technique to a wide variety of biological systems.

Geoff Howlett, who officially retired in 2012 but retains an honorary appointment in the department, has studied the misfolding and aggregation of plasma apolipoproteins and their propensity to accumulate in disease-related amyloid deposits. His work aims to discover the mechanism that drives normal proteins to form amyloid fibrils and to develop inhibitors that block and reverse fibril growth. This is of significance in research into a number of age-related diseases including dementia, diabetes and heart disease. The techniques used by the group include analytical ultracentrifugation, fluorescence, NMR and mass spectroscopy to explore the formation and breakdown of amyloid fibrils and the interactions of fibrils with other amyloid components.

One of the current research themes in the department is cell signalling. This research was initially established by Heung-Chin Cheng (1958–), who came to the University of Melbourne in 1992 from North America. He had taken his BSc in 1981 from the Chinese University of Hong Kong and his PhD in endocrinology from the University of California, Davis in 1986. From 1987 to 1988 he was a postdoctoral fellow in the Laboratory of Molecular Endocrinology at the University of California, San Francisco, and from 1989 to 1992 in the MRC Group in Signal Transduction in the Department of Medical Biochemistry at the University of Calgary, Alberta.

Heung-Chin Cheng heads a laboratory that forms part of the group of three teams in Bio21 dedicated to signalling

One of his most-cited papers is based on his PhD project.[9] Defining the determinants in PKA inhibitor governing specific and potent inhibition of PKA, the investigators' results facilitated determination of the first crystal structure of a protein kinase.

Cheng came to the University of Melbourne in 1992 as a lecturer, was promoted to senior lecturer in 1997 and associate professor in 2007. His laboratory forms part of the group of three teams in Bio21 dedicated to signalling. On joining the Melbourne department, Cheng continued his investigations into how Src-family kinases are regulated by their endogenous inhibitors CSK and CHK. His group identified two new auto phosphorylation sites in the kinases and discovered for the first time that CHK employed a unique non-catalytic mechanism to inhibit Src-family kinase activity. He moved his research group, which studies signal transduction enzymes associated with cancer and neuronal loss in acute brain injury such as stroke and neurodegenerative diseases such as Parkinson's disease, to Bio21 in 2005. In 2013 Cheng and his colleagues published a report that details a new mechanism of neuronal death in ischemic stroke condition and reveals for the first time that calpain cleavage can convert the protooncogenic protein kinase Src from a promoter of neuronal survival to a mediator of neuronal death.[10]

Another research strength of the department is in the area of host–pathogen interactions, with a focus on parasitic infections. There are currently four laboratories actively involved in understanding the molecular biology of the interactions between the host and the pathogen. This field of research was pioneered in the department by Malcolm John McConville (1958–), who joined it in 1994 as a Wellcome Trust senior research fellow after five years in Scotland. McConville, who took his BSc in 1980 and his PhD on the ecology, metabolism and polysaccharide structure of Antarctic sea ice diatoms in 1985, took up a position as research officer at the WEHI in 1986 before leaving for the Department of Biochemistry at the University of Dundee. There, he was successively a Royal Society Florey fellow (1989), C.J. Martin research fellow (1990 to 1991) and Wellcome Trust senior research fellow from 1992 until he returned to Australia in 1994, holding an Australian Wellcome Trust Senior Research Fellowship until 1998. In 2010 McConville was appointed deputy director of the Bio21 Institute and associate director of the Structural Biology Research Theme. His longstanding interest is in the metabolism of microbial pathogens with a view to identifying new drug targets. His research is directed towards understanding how microbial pathogens survive in humans and other animal hosts.

McConville's team's research focuses on the eukaryotic microbial pathogens that cause a number of important and neglected human diseases. These include *Plasmodium falciparum*, the cause of malaria; *Toxoplasma gondii*, a protozoan parasite that causes human toxoplasmosis; and *Leishmania spp*, human leishmaniasis. There are no vaccines against any of these diseases

Malcolm McConville works on microbial pathogens, including those that cause malaria, human toxoplasmosis and human leishmaniasis

and current drug treatments are both limited and constantly being undermined because the parasites become drug-resistant. Identifying and validating new drug targets requires a deeper understanding of the biology and metabolism of these pathogens in vivo. In 2005, the team announced that it had found that the leishmaniasis parasite does not use glucose for energy storage, but rather uses mannose, a different sugar. This discovery may help in the development of drugs for many microbial pathogens which use mannose, including those involved in malaria and tuberculosis.

The group uses a range of approaches in identifying the parasite metabolic pathways and host responses that parasites need in order to survive in their human and animal hosts, including comprehensive metabolite profiling or metabolomics, the non-targeted detection and quantification of small molecules or metabolites in biological materials such as plasma, urine, tissue, plant and microbial extracts. The Victorian node of Metabolomics Australia, located at the School of Botany (Victorian and Australian Centre for Plant Functional Genomics) and the Bio21 Institute for Molecular Science and Biotechnology, offers an advanced analytical facility providing state-of-the-art metabolomics infrastructure, including access to expertise and technologies that cover a wide range of metabolite chemistries and quantitative analyses required for comprehensive metabolite profiling applicable to biomedical, agri-food and environmental sciences.

Another group is dedicated to solving the structure of proteins. Dick Wettenhall built up strength in NMR structural biology through the recruitment of Paul Raymond Gooley (1955–), who joined the department in 1996 after spending 1987 to 1988 as a postdoctoral scientist at Yale, 1988 to 1991 at the University of Arizona; and 1992 to 1996 as a senior research chemist/research fellow at Merck and Co. in New Jersey. It was during that last period that he produced what he regards as his most significant scholarly publications.

Paul Gooley heads a laboratory working on nuclear magnetic resonance spectroscopy (protein structure and function)

Three papers, written while he was at Merck, were produced during the development period of triple resonance NMR spectroscopy as a structural biology tool. For a moment in time, somewhere between 1995 and 1996, the protein was the largest solved by NMR experiments.[11] Gooley had gone to America to pursue his postdoctoral work after graduating BSc (Hons) in 1982 and taking his PhD in 1985 from the University of New South Wales with a thesis on 1H NMR studies of cardiostimulant polypeptides from sea anemones.

Paul Gooley first joined the department as a senior lecturer, becoming an associate professor in 2006; he was Deputy Head of Department from 2006 to 2013. The projects on which the Gooley group has worked include structural and dynamic investigations of protein–protein, protein–peptide and protein–lipid interactions, including the relaxin family of receptors which are GPCRs, their interactions with their respective hormones and their curious mechanisms of activation; the specificity and affinity for oligosaccharides by the carbohydrate binding modules, and their significance in regulation of the enzyme AMP protein activated kinas, protein import receptors of mitochondria, and how they bind and pass preproteins through the membrane; and apolipoprotein CII in the amyloid state from lipid and lipid-free environments. The group is also interested in applying NMR methods to metabolomics, focusing on the role and impact of the gut microflora in health and in syndromes such as chronic fatigue. They are also studying the effect of prebiotics and pro-biotics on the exometabolome of the microbiota. Gooley, together with Terry Mulhern, provide expertise to the department in structural biology using a range of NMR approaches.

Like Brent Smith, Paul Gooley is an enthusiastic cyclist who has participated in several Around the Bay rides. This annual event, which raises money for charity, was first held in 1993. It initially consisted of a 210-kilometre route around Port Phillip Bay, from Melbourne to Queenscliff via the Westgate Bridge, with a ferry ride to Sorrento, then back up to Melbourne again via Beach Road, though there have been minor changes since then. The 2012 event attracted 17000 participants from Australia and overseas. Gooley has twice participated in the Great Ocean & Otway Classic Ride, which begins and ends at Elephant Walk Reserve on the foreshore at Torquay and travels through Torquay, Moriac and Deans Marsh to Lorne. The return route takes the riders through Fairhaven, Aireys Inlet and Anglesea.

Paul Gleeson and Ian van Driel were recruited as laboratory heads from Monash late in 2001. Gleeson has been Professor and Chairman of both the Department of Biochemistry and Molecular Biology and the Bio21 Molecular Science and Biotechnology Institute since 2006. The Gleeson and van Driel groups, who had already been actively collabor-ating for more than a decade, expanded expertise in the department in molecular cell biology, molecular immunology and chronic inflammatory diseases, as well as emphasising a team approach to establishing multidis-cipline research programs. They were excited by the potential of Bio21 and the range of emerging platforms, as well as by the potential oppor-tunities for the department in both the recruitment of new people and generational change. Their large groups were initially located in a joint facility at one end of the first floor in the Russell Grimwade Building, the old lab of Barrie Davidson.

Paul Anthony Gleeson (1952–) was awarded his BSc (Hons) at Melbourne in 1973 in the Department of Microbiology and his PhD in 1980 in the School of Botany. His doctoral thesis was on the characterisation of carbohydrate polymers associated with pollination in gladiolus and secale. He spent 1980 to 1982 as a postdoctoral fellow in the Department of Biochemistry of the Hospital for Sick Children in Toronto, identifying new Golgi enzymes in the glycosylation pathway. Having won a Beit Memorial Medical Research Fellowship in 1982, he spent the next two years in London in the Division of Biochemistry of the National Institute for Medical Research, studying cell surface carbohydrates and cell–cell adhesion, before taking up a position as research fellow in the Department of Biochemistry at La Trobe University. In 1986 he accepted a lectureship in the Department of Pathology and Immunology at Monash University, attaining the rank of associate professor before leaving to join the Department of Biochemistry and Molecular Biology at the University of Melbourne in late 2001. He also spent periods of time at the European Molecular Laboratory in Heidelberg (1992) and at the Institut Curie in Paris (2005), as a recipient of a Rothschild Award.

Gleeson's research interests lie in the molecular basis of organ-specific autoimmune diseases and the molecular mechanisms of intracellular membrane transport. His research dissects complex cell interactions of

Paul Gleeson, seen here with his research assistant Fiona Houghton, has been Head of Department since 2006

the immune system. He and Ian van Driel have established one of the best-defined mouse models for investigating loss of immune tolerance to self-antigens and the development of organ-specific autoimmune diseases. Gleeson's research also applies a range of cell biological approaches designed to discover the molecular machines that regulate transport pathways in a range of different systems, including transformed and primary cells and whole organisms. His team is particularly interested in the regulation of the cargo transport from the Golgi apparatus to the surface of the cell, and the recycling of cargo through endosomes back to the cell surface, pathways relevant to many normal physiological systems, such as secretion of cytokines by immune cells, and also relevant to the production of the cytotoxic amyloid peptide in Alzheimer's, a topic under investigation in his laboratory. His work on understanding the molecular mechanism of membrane transport has led to collaborations with the biotechnology industry. In 2013 Gleeson, Anne Verhagen of the WEHI and Steven Downer of CSL Limited were successful in obtaining a four-year ARC Linkage grant for their project investigating intracellular trafficking and function of a recycling receptor which prolongs the serum half-life of novel therapeutic proteins. This project will exploit interactions with a natural receptor, which prolongs the life span of serum proteins, to enhance survival of therapeutic engineered proteins. The life span of recombinant engineered proteins for therapeutic use is a critical factor in their effectiveness, ease of clinical application and cost.

Paul Gleeson enjoys a variety of physical challenges, in particular bushwalking and cycling, as well as escaping to his holiday house at Waratah Bay in Gippsland and sharing the delights of Wilson's Promontory with colleagues and friends.

Ian Richard van Driel (1960–) came to the department from Monash in the same year as Paul Gleeson. Having taken his BSc (Hons) from the University of Western Australia in 1981, he was awarded his PhD from the WEHI in 1986 for his thesis on the structures of the murine plasma cell antigen PC-1 and the receptor for tranferrin. He spent 1986 to 1988 with a C.J. Martin Fellowship in the laboratories of the Nobel Prize–winning scientists Joseph Goldstein and Michael Brown at the University of Texas Southwestern Medical Center, where he was a postdoctoral research fellow and assistant instructor in the Department of Molecular Genetics. Returning to Australia, van Driel took up an appointment in the Department of Pathology and Immunology in the Monash University Medical School in 1989, rising to associate professor before transferring to the Melbourne Department of Biochemistry and Molecular Biology in 2001. He was appointed professor in 2011.

The van Driel laboratory investigates various aspects of immunology and inflammatory disease, including immunological tolerance, or how the immune system learns to tolerate our own tissues while still maintaining the ability to attack foreign invaders; autoimmune diseases, by examining their molecular and cellular causes; and inflammatory bowel disease, by

Ian van Driel's laboratory investigates various aspects of immunology and inflammatory disease

discovering new models of IBD and finding the genes responsible. The group's research into functional genomics of autoimmunity (discovering genes that predispose to autoimmunity) may allow new predictive and therapeutic approaches. Much of van Driel's work focuses upon understanding the events associated with immunological tolerance and autoimmune disease, using gastric autoimmunity as a model system. His laboratory also investigates the cell biological events that control the secretion of stomach acid. An early paper with colleagues from the WEHI, the Ludwig Institute for Cancer Research and the Monash Medical School published in *PNAS* was the first report of the isolation of a gene from a plasma cell protein and one of the first to report the cloning of a gene for a membrane protein.[12] More recent work involved a project to discover genes that are implicated in gastrointestinal disease by screening collections of mutant mice.[13]

The research theme of protein misfolding in the department was significantly enhanced by the recruitment of a new staff member in 2002.

When Andrew (Andy) Francis Hill (1970–) moved into his quarters in the Russell Grimwade Building in 2002, he was told that he was breaking a tradition established by Bruce Grant and Bruce Livett in occupying a space previously occupied only by laboratory heads called Bruce. His was the first group to move in 2005 into the Bio21 Institute. Andy Hill took his BSc (Hons) from Victoria University, Wellington before leaving New Zealand for the country he had left at the age of five, to work as a research assistant in the laboratory of John Collinge in the Prion Disease Group at St Mary's Hospital Medical School in London. The following year, in 1993, he began working towards his PhD researching human prion diseases, such as Creutzfeldt-Jakob disease (CJD). When a new form of CJD was discovered (variant CJD) that occurred in much younger people and was linked with exposure to an animal prion disease – bovine spongiform encephalopathy (BSE, or 'mad-cow' disease) – in 1996, this work turned to investigating the biochemistry of prion proteins in patients with this new disease. The research revealed a molecular link between BSE and vCJD and the team also developed a diagnostic test based on sampling

Andrew Hill, seen here with research assistant Robyn Sharples, investigates processing of proteins and RNA in neurodegenerative diseases

tonsil biopsies. This work, which had the effect of delaying completion of Hill's PhD, of which it formed the basis, was published in several papers in *Nature* and *Lancet*. 'Molecular Analysis of Prion Strain Variation and the Aetiology of "New Variant" CJD' has been cited over 1500 times.[14] Hill took his DIC from Imperial College London in the same year as he was awarded his PhD, 'Molecular Studies on Human Prion Proteins', and remained with the group for another eighteen months to complete some long-term experiments that involved keeping mice for their entire life span of between two and three years. The finding that prions can 'hide' in the body had implications for public health.

In 2002, having spent time with the Prion Group and with Colin Masters working on neurodegenerative diseases like Alzheimer's and Parkinson's, he returned to Melbourne, joining an NHMRC Program Grant as chief investigator as well as bringing Project Grant funding of his own. He accepted a senior lectureship in the department at the end of 2002 and, having received a NHMRC R.D. Wright Career Development Award, was able to continue as a research fellow. Andy Hill won a succession of awards, including a Young Investigator Award from the International Society for Neurochemistry in 2004 and 2007, a Young Investigator Award from the Lorne Conference on Protein Structure and Function in 2004, the Edman Award in 2005, a Victorian Young Tall Poppy Award in 2006 and the Merck Research Excellence Medal from the Australian Society of Biochemistry and Molecular Biology in 2010. He was promoted to associate professor in 2008 and professor in 2012. Andy Hill has been on competitive fellowships continuously since 1999, but should these fail or should he wish to change his career, Melbourne is probably the ideal city for a man who is also a qualified barista.

Protein misfolding and neurodegeneration in the department was further strengthened by the return to Melbourne of Danny Hatters in 2007 with a C.R. Roper Fellowship, followed by the Grimwade Research Fellowship in 2009.

Another 2007 recruit to the department enhanced the cell signalling theme. Marie Ann Bogoyevitch (1965–) graduated BSc (Hons and a university medallist) in 1986 and took her PhD in 1990 from the University of Queensland. On the occasion of taking up the position of secretary of the Australian Society of Biochemistry and Molecular Biology in 2007, she wrote of her early fascination with biochemistry, at a time when investigation of signal transduction was just beginning.[15] Marie Bogoyevitch completed her PhD in record time because, having applied for a position in the National Heart and Lung Institute in London in 1989, success meant starting early the following year. She spent five years in the Department of Cardiac Medicine of the institute, initially as a postdoctoral fellow and later as a lecturer. After investigating cAMP-dependent signalling then protein kinase C isoforms in the heart, her work turned to research into the

Marie Bogoyevitch leads a research team investigating signal transduction

mitogen-activated protein kinases and their upstream regulators. From 1995 to 1997 she worked at the Cancer Research Campaign Centre for Cell and Molecular Biology, Chester Beatty Laboratories at the Institute of Cancer Research in London, her focus switching to more fundamental issues in signal transduction.

As Bogoyevitch told *Australian Biochemist*:

> I found myself flicking through an issue of *Nature*, when I chanced upon an advertisement for a Lectureship at the University of Western Australia (UWA). I think they say, 'the rest is history'. After a whirl-wind trip to Perth, I was happy to accept the position and return to Australia in mid-1997 to establish my own laboratory.[16]

Over the next decade, Bogoyevitch rose from lecturer to associate professor, noting the inevitable strains of juggling coordination of the work of her own research team, teaching undergraduate students and increasing involvement in university administration. During this time she led the Cell Signalling Laboratory within the School of Biomedical Biomolecular and Chemical Sciences and won a UWA Teaching Excellence Award.

At the University of Melbourne, where she moved in mid-1997 as associate professor, Marie Bogoyevitch leads a research team investigating signal transduction, which is the unravelling mechanisms of intracellular communication pathways in both health and disease. The investigators are exploring the regulation of intracellular communicators that provide critical control points during signal transduction events. A specific focus lies in the c-Jun N-terminal Kinase (JNK) subfamily. These enzymes have attracted increasing interest following their initial description as stress-and-cytokine-activated protein kinases and are now implicated as mediators in diseases including stroke, obesity and diabetes. Although it is hoped that greater understanding of these kinases will facilitate the development of improved therapeutic strategies for such diseases, controversy continues to surround the biological functions of the JNKs. Marie Bogoyevitch's review of JNK substrates, the first definitive look at the proteins targeted by these kinases, which built on the work of the Bogoyevitch laboratory exploring the molecular concepts of substrate recognition, has now been widely cited.[17]

One of the senior members of the Marie Bogoyevitch laboratory who came to the department with her from Western Australia subsequently developed his own research projects and established his own lab in 2012. Dominic Chi Hiung Ng (1977–) had taken his BSc (Hons) in 1999 and his PhD – on characterising intracellular signalling mechanisms involved in the progression of cardiac hypertrophy and failure – in 2004 from the University of Western Australia. As an undergraduate he had won the Swan Brewery Prize in Biochemistry and the Lugg Medal in Biochemistry. In 2007 he received the inaugural Bendat Family Foundation Fellowship Award from the National Heart Foundation. This award, presented for outstanding cardiac

Dominic Ng won the 2011 ASBMB Edman award and is currently working on microtubule dynamics

research by a young Western Australian, is an initiative of the noted philanthropist Jack Bendat, who migrated from the United States in 1966 and made his fortune in shopping centres and other ventures such as Goundrey Wines, and also has been the owner of the Perth Wildcats basketball team since 2006. In 2011 Ng won both the Bioplatforms Australia Award and the ASBMB Life Technologies Edman Award. He received an ARC Future Fellowship in 2013. Before joining the department in 2007, Ng worked principally in Western Australia, with a two-year period in Singapore as research fellow at the Institute of Molecular and Cell Biology, part of the Agency for Science, Technology and Research.

Also joining the department in 2007 as a member of the host–pathogen interactions research theme was Stuart Ralph. This represented his return to the University of Melbourne, where he had earlier completed his BSc (Hons) and then his PhD in 2003. This work, supervised by Geoff McFadden in the School of Botany, studied the function of the relic chloroplast (called an apicoplast) in the malaria parasite *Plasmodium falciparum*. Ralph was awarded a NHMRC C.J. Martin Fellowship to support his postdoctoral research periods in France and Australia. The first of these was with Artur Scherf at the Institut Pasteur in Paris, where he

studied the mechanism by which malaria parasites switch their surface proteins to avoid destruction by the human immune system. The second, with Alan Cowman at the Walter and Eliza Hall Institute of Medical Research in Melbourne, focused on the characterisation of drug resistance and drug targets in malaria.

Ralph was appointed as a group leader in the Department of Biochemistry and Molecular Biology in 2007, supported initially by a C.R. Roper Fellowship and Biochemistry Fund Fellowship and subsequently by an ARC Future

Stuart Ralph is interested in parasitic diseases, with a primary focus on the causative agent of severe malaria, *Plasmodium falciparum*

Fellowship and an NHMRC R.D. Wright Career Development Fellowship. Here, Ralph extended his work on malaria and other neglected tropical diseases, focusing particularly on the mechanism of protein translation within the parasite. This research combined bioinformatic approaches to identify and prioritise drug targets with molecular cell biology and biochemistry to characterise and validate those proteins. This period coincided with the sequencing of the genomes and transcriptomes of many parasites, and Ralph was extensively involved with the analysis and publication of several of these genomes. He collaborated widely with international parasite biologists and bioinformaticians to further this work on parasites, first through a World Health Organization–sponsored group to develop tools to prioritise drug targets for neglected tropical diseases, and then through a European Union initiative to develop drugs against parasite protein translation. Ralph's research is very highly cited and has been published in leading international journals such as *Cell*, *Science*, *Nature* and *PNAS*, with cover images in both *Nature* and *Cell*.

Frustrated by the lack of support for anti-infective drug development from many major pharmaceutical companies, Ralph became increasingly active in alternative strategies for drug innovation. These principally involved open-source approaches that employ collaborative systems to engage multiple international teams and individuals in drug research. Ralph participated in drug discovery projects with European and American researchers working on tRNA synthetase, as well as with an Australian-led international open-source project to improve anti-malarial drugs. He also joined the largest of these international initiatives, the Indian Open Source Drug Discovery project. Ralph was awarded a Churchill Fellowship to spend several months in India in 2012–13, building links with Indian parasite and drug discovery teams.

Stuart Ralph remains committed to the development of fundamental parasite research and its exploitation to discover opportunities for human therapeutic interventions. He has taught widely in international parasitology courses, and in 2014 helped to establish a parasite biology course for the Australian Society for Parasitology.

The person working in the area of protein folding diseases to most recently set up his own laboratory (in 2013) joined the department in 2006, initially working in Geoffrey Howlett's team. Michael Donald

Michael Griffin heads a laboratory that is interested in understanding both functional and misfunctional protein–protein interactions

William Griffin (1977–) took both his BSc and PhD from the University of Canterbury, New Zealand, in 1999 and 2005 respectively. His thesis 'Why Is Dihydrodipicolinate Synthase a Tetramer?', supervised by Juliet Gerrard, explored the functional consequences of protein–protein association in dihydrodipicolinate synthase (DHDPS). During his graduate studies, with the support of both a PhD Scholarship and a travel award from Crop and Food Research NZ Ltd that enabled him to attend the Fifth European Symposium of the Protein Society, Griffin gained extensive expertise in enzymology, biophysics and structural biology, and contributed greatly to the understanding of DHDPS structure and function.

Working with Howlett, Griffin focused on aberrant protein misfolding resulting in amyloid fibril formation. This work has led to a number of critical contributions to the field of amyloid research, the foremost of these demonstrating that methionine oxidation causes apolipoprotein A-I to misfold and form amyloid fibrils (MG6). This is the first such demonstration and represents an important advance in the understanding of apoA-I amyloidosis. The findings also have significant implications in the wider amyloid field, as well as in research on apolipoproteins, lipid transport and other facets of cardiovascular disease. The work was published in *PNAS* with Griffin as senior and corresponding author.[18] He was awarded a Melbourne Early Career Researcher Grant in 2009 to complete and extend this work, and received the Young Investigator Award at the thirty-fifth Lorne Conference on Protein Structure and Function after presenting a lecture on this research.

Michael Griffin won an Australian Research Council Post Doctoral Fellowship in 2011 and the 2013 C.R. Roper Fellowship in 2013, the year in which he established his own laboratory. He has led the Bio21 Institute Australian Synchrotron Crystallography Users Group and is principal investigator on Bio21 applications for Australian Synchrotron Crystallography beam-time. He is also the chief investigator and University of Melbourne node manager of an ARC LIEF grant for the MATRIX high-throughput X-ray facility (a joint facility with La Trobe University). In this role he was responsible for the installation of the first protein X-ray diffractometer for protein crystallography within the Bio21 Institute. He also chairs the Bio21 Macromolecular Interactions Facility Committee, which manages specialised biophysical instrumentation.

The most recent laboratory head to join the host–parasite research theme was, like Malcolm McConville, returning to her alma mater. Leann Tilley (1958–) had taken her BSc (Hons) from the University of Melbourne in 1978, where she worked as a research assistant from 1979 to 1980 before moving interstate for her graduate studies. Her PhD from Sydney University was awarded in 1985. She spent a year at Utrecht University in the Netherlands as a postdoctoral fellow, where she characterised the ATP-dependent 'flippase' responsible for aminophospholipid asymmetry in red blood cells. This pioneering study sparked much more work on the role of phosphatidylserine asymmetry in apoptosis. In 1986 she took up a Poste Orange Postdoctoral Fellowship offered by the Institut National de la Santé et de la Recherche Médicale, working at the Collège de France. On her return to Australia, Tilley spent two years as a postdoctoral fellow at the University of Melbourne before taking up a lectureship at La Trobe, where, having been promoted to professor in 2004, she remained until her return to the Department of Biochemistry and Molecular Biology at Melbourne in July 2011.

As well as being the recipient of numerous awards, including the Bancroft-Mackerras Medal from the Australian Society for Parasitology in 2010 and the Beckman Coulter Science Discovery Award in 2011, Leann Tilley, who is Deputy Head of the Department, is one of the few women to have received an Australian Research Council Professorial Fellowship. She has played a leadership role in the ARC Centre of Excellence for Coherent X-ray Science as

Leann Tilley heads a laboratory that is working as part of a global effort to understand and control malaria

its deputy director from 2006 to 2012 and as director from 2013. Her work embraces technologies ranging from drug and protein chemistry to molecular cell biology and novel imaging technologies. She works in collaboration with experts from other disciplines, ranging from molecular parasitologists to organic chemists and optical physicists. The Centre of Excellence for Coherent X-ray Science brings together physicists, chemists and biologists to develop fundamentally new approaches to probing biological structures and processes. It has received international acclaim for its cross-disciplinary and cross-institutional work, and for its contributions to the development of novel imaging techniques. The Tilley laboratory has helped to develop and implement a number of new imaging modalities, including coherent X-ray diffraction imaging, 3D electron tomography, cryo X-ray tomography and structured illumination microscopy, and to apply them in pioneering applications.

The Tilley laboratory works as part of a global effort to understand and control malaria, a disease that kills over half a million people each year, undertaking research in the areas of cell biology and drug development related to the malaria parasite, *Plasmodium falciparum*. Its investigations focus particularly on how the parasite alters the erythrocyte surface to cause malaria pathology, as well as the remarkable transformation that makes parasites become banana-shaped and allows them to be transmitted from a human host to a mosquito vector. The team is also investigating the action of and resistance to artemisinin, with a view to designing better antimalarial drugs. Among Leann Tilley's many widely cited papers on malaria parasites, one on the molecular basis of artemisinin resistance was highlighted in *PNAS New & Newsworthy* and *PNAS Commentary*.[19] The team notes that reports of emerging resistance to first-line artemisinin antimalarials make it critical to define resistance mechanisms. Tilley's work demonstrated that parasites can shut down their growth when exposed to the drug. As artemisinins only last a few hours in the bloodstream, parasites that do this more efficiently would be better able to survive until the drug reaches sub-lethal concentrations.

In a pleasing link with the department's past, Leann Tilley enjoys the occasional holiday in her beach house at Wye River – a house that was built by Audrey Cahn, whom she interviewed in 1998 and whose obituary she later wrote.[20]

The most recently arrived laboratory head to join the molecular immunology theme came to Australia in 1998, working with Ken Shortman in the WEHI, where he established his own laboratory in 2001. He came to the University of Melbourne in 2011. José Alberto Villadangos (1965–), who was born in Spain, took his MSc in 1989 and his PhD in 1994 from the Universidad Autónoma de Madrid, after which he pursued his postdoctoral research in the United States with Hidde Ploegh (1953–), first from 1994 to 1997 at the Center for Cancer Research of the Massachusetts Institute of Technology and then from 1997 to 1998 in the Department of Pathology of Harvard Medical School. He came to Melbourne with a Human Frontiers Science Program Fellowship. This international program funds research on the complex mechanisms of living organisms. Villadangos joined the Immunology Division of the WEHI and in 2000 was made a special fellow of the Leukemia and Lymphoma Society, an award intended to permit the recipient to establish an independent research program. Villadangos started his own group in 2001. From 2004 to 2009 he was named a Leukemia and Lymphoma

Society scholar, which designates a highly qualified investigator who has demonstrated a capacity for independent, sustained and original investigation in the field of leukaemia, lymphoma and myeloma. The recipient must hold an independent faculty-level or equivalent position and have obtained substantial support for his or her research from a national agency. From 2009 to 2011, José Villadangos was an NHMRC senior research fellow at the WEHI and an honorary senior fellow in the Department of Medical Biology at the University of Melbourne.

José Villadangos holds a dual appointment in the Departments of Biochemistry and Molecular Biology and Microbiology and Immunology

In his current position as professor at the University of Melbourne, Villadangos holds a dual appointment in the Department of Biochemistry and Molecular Biology, and the Department of Microbiology and Immunology. Villadangos's current work examines the complexity of the dendritic cell network in vivo and the mechanisms by which immune cells capture, process and present antigens. His current research interests include the mechanisms of antigen presentation in dendritic cells, the regulation of proteostasis by membrane ubiquitin ligases, the mechanisms controlling the dimerisation of the amyloidogenic protease inhibitor Cystatin C, autophagy, and the development of adoptive cell therapy for the treatment of cancer. One of his most-cited papers deals with cross-presentation, dendritic cell subsets and the generation of immunity to cellular antigens. Another, which attracted the attention of the *Herald Sun*, ABC Radio in Adelaide and numerous online news sites, describes the role of a component of the endocytic route, IFITM3, in the survival of tissue-resident memory T cells (Trm) during viral infections. It represents his team's successful extension to T cell biology of the approach they have previously followed to characterise DC function at the biochemical/cell biological level.[21]

In 2013, in collaboration with Monash University, Deakin University and the Macfarlane Burnet Centre for Medical Research, Villadangos was awarded a Linkage Infrastructure Equipment and Facilities Grant for an advanced in vivo imaging facility. This project will establish an advanced in vivo imaging facility for examining host–microbe interactions and associated immunological processes within the context of the numerous infectious disease models within the University of Melbourne and associated collaborators. The Zeiss LSM 7MP 2-photon imaging system will provide enhanced capacity to directly visualise cellular and molecular events in real time, with greater sensitivity and in a broader range of tissues and organs. This will provide the opportunity for novel insights into numerous immunological and host–microbe interactions.

What follows is a brief account of the work of two women who took up positions as laboratory leaders in 2012 and 2013, respectively. Kathryn Elizabeth Holt took her BA in 2003 and her BSc (Hons) in 2004 from the University of Western Australia, majoring in biochemistry, applied statistics and philosophy. Her PhD in molecular biology from the Wellcome Trust Sanger Institute and the University of Cambridge focused on pathogen genome sequencing.[22] On her return to Australia in 2009, Holt took up an NHMRC Postdoctoral Fellowship in the Department of Microbiology and Immunology, and while working as a research fellow she undertook a Masters in Epidemiology at the university, which she was awarded in 2011. She was promoted to senior research fellow in 2014, the same year in which she was awarded the highest-ranked Career Development Fellowship – Biomedical, Level 1 (for applicants who have completed their PhD in the last two to seven years) by the NHMRC.

Kathryn Holt was awarded a L'Oréal For Women in Science Fellowship in 2013 and investigates pathogen genomics and bioinformatics, bacterial populations and bacterial communities

The following year, she was awarded two international travel grants. One, to present at a conference in the United States, was from the private philanthropic CASS Foundation, which was established in 2001 to support and promote the advancement of education, science and medicine, and research and practice in those fields. The other was a Ron Rickards Fellowship from the Australian Academy of Science, which permitted her to work with François-Xavier Weill at the Institut Pasteur. There, they used high-throughput genomics to study the global emergence of a highly drug-resistant form of *Salmonella* associated with the consumption of chicken meat. This genomic data was able to explain the abilities of the pathogen to survive in the presence of a wide range of antimicrobial drugs, and showed that this highly adaptive bacterium has gained these abilities many times in different countries, in response to different patterns of antimicrobial usage. Holt and her hosts also began a collaboration to investigate the emergence and global spread of *Shigella dysenteriae*, the key cause of severe drug-resistant dysentery epidemics which have swept through refugee camps in Asia and Africa.

Kathryn Holt moved to the Bio21 Institute at the University of Melbourne shortly after her return to Australia and started her own genomics laboratory. She was awarded four NHMRC research grants that will use genomics in a variety of ways to study human infections. In 2013 she was one of two Melbourne University medical researchers awarded a L'Oréal for Women in Science Fellowship, which she will use to understand how antibiotic-resistant bacteria spread in Melbourne hospitals. One of her major projects is using genomics to study the origin, evolution and spread of antibiotic resistance in hospitals, particularly the role of *Klebsiella*, a bacterium that is rapidly becoming a problem worldwide because of its capacity for survival in a broad range of environments and its propensity for picking up novel genes.

Essentially Kathryn Holt's research involves genetics, mathematics and the power of supercomputers to study the genome of deadly bacteria. The field of bioinformatics gained special prominence in Australia in 2013 when Terry Speed (1943–) of the WEHI, with whom Holt worked, was awarded that year's Prime Minister's Prize for Science for his work using mathematics and statistics to help biologists understand human health and disease.

Diana Stojanovski (1980–), who joined the molecular cell biology theme of the department in January 2013, took both her BSc (Hons) and her PhD, which she was awarded in 2006, from La Trobe University. Stojanovski spent from October 2005 to the beginning of 2009 in Germany working with Nikolaus Pfanner (winner, with Jürgen Soll, of the 2004 Gottfried-Wilhelm-Leibniz Prize) at the Institute for Biochemistry and Molecular Biology of the Albert-Ludwigs-Universität in Freiburg, while on an Alexander von Humboldt Research Fellowship. About 600 such fellowships are awarded worldwide every year.

Diana Stojanovski runs the Mitochondrial Biogenesis and Disease Laboratory and won the ASBMB Edman Award in 2013

They are available only to an experienced researcher less than twelve years after completing their doctorate. The fellowships allow holders to carry out a research project of their choosing, lasting between six and eighteen months, in cooperation with an academic host of their choice, at a research institution in Germany.

Returning to La Trobe in 2009, Stojanovski spent ten months as a postdoctoral fellow in the ARC Centre of Excellence for Coherent X-Ray Science, subsequently becoming research fellow and group leader in the La Trobe Institute for Molecular Science, where she set up an independent research group dealing with mitochondrial protein trafficking. She joined the Department of Biochemistry and Molecular Biology as the Biochemistry Fund fellow and runs the Mitochondrial Biogenesis and Disease Laboratory. Mitochondrial function extends well beyond that of energy generation, because mitochondria serve as sensors for multiple metabolic functions and are key players in health and disease. These functions are intimately dependent on the protein complement of the organelle. Mitochondria cannot be created de novo and therefore require the constant synthesis of mitochondrial- and nuclear-encoded proteins for their biogenesis. Disturbances in mitochondrial protein homeostasis compromise organelle function. Mitochondrial dysfunction is associated with numerous disease states, including cancer, Alzheimer's and Parkinson's. Research in the Stojanovski laboratory is directed towards understanding the machineries and mechanism that drive protein import and assembly in mitochondria, and unravelling the link between defects in mitochondrial protein import and assembly and human disease.

Diana Stojanovski's many awards include receipt of the Lorne Young Investigator Award in 2009, both the La Trobe University Dean's Excellence in Research Award for Early Career Researchers and Finalist for the Scopus Young Researcher of the Year Award in 2010, an EMBO Short Term Fellowship in 2011, a CASS Travel Grant in 2012 and the ASBMB Edman Award in 2013.

The work of seventeen people heading laboratory teams in the Bio21 Molecular Science and Biology Unit has been briefly described above. Together, as well as leading research in their fields, they are responsible for training the postgraduate students in biochemistry and molecular biology. They also give lectures to undergraduate students in biochemistry.

Staff in the atrium of the Bio21 Institute, 2014

CHAPTER 11

A Bridging Science

T HE SCIENCE OF biochemistry has undergone fundamental changes at the University of Melbourne since its early days there, just as it has been transformed all over the world. The concerns of Biochemistry at Melbourne have altered and developed, and the department's physical environment has also changed. The emergence of molecular biology, combining aspects of biochemistry, genetics, microbiology, virology and physics, roughly coincides with the establishment in 1938 of the Department of Biochemistry, the transformation mirrored in its later designation as the Department of Biochemistry and Molecular Biology.

The links between the study of biochemistry and chemistry at the University of Melbourne have always been strong, from the earliest times, and several biochemists, notably William Alexander Osborne and Arthur Rothera from the Department of Physiology, addressed meetings of the Melbourne University Chemical Society before 1938. Paul Eirich and Bill Rawlinson gave papers to the society in 1942 and 1943, while Victor Trikojus spoke in 1944, 1945, 1947 and 1952. In 1946 the society heard Joseph Lugg; Neil Izaacs and Richard Pau addressed it in 1978 and 1999. The decline in participation by members of the Department of Biochemistry in the proceedings of the Melbourne University Chemical

Society since the 1950s may be the result of closer links established with other organisations such as the Florey, the WEHI and CSIRO.

Many members of the department, however, have undertaken joint research with members of the Department of Chemistry, including Frederick Collins, Bruce Grant, W.H. Sawyer, Pamela Todd and Trikojus himself. Arthur Rothera was a member of the Faculty of Agriculture from his arrival in 1906. Francis Hird held a dual appointment in the Department of Biochemistry and the Faculty of Agriculture. Others have held dual appointments in Biochemistry and the Department of Microbiology and Immunology. Biochemistry is a vital component in the work of the university's botanists, with Adrienne Clarke moving from the department to a career in the School of Botany, and John Chippindall, while still in the Department of Biochemistry, designing equipment for use there. It also plays a vital part in the work of veterinary scientists. Biochemistry is indeed a science that links and forms a core part of many disciplines.

Throughout its history, however, the Department of Biochemistry and Molecular Biology has always been part of the Faculty of Medicine, Dentistry and Health Sciences. The earliest biochemists at the University of Melbourne were predominantly medical doctors, though medicine was often not their first career choice. W.A. Osborne undertook his medical studies reluctantly and 'escaped into science'. Charles Martin,

Terrence Mulhern, Heung-Chin Cheng, Paul Gooley, Ken Gayler and Bruce Livett

who qualified as a medical practitioner, spent his life in research in physiology, apart from a period in the Australian Army Medical Corps in the Middle East and France as a pathologist. When World War II was declared, however, he had part of his own property, Roebuck House, fitted up as a biochemical laboratory.[1] Arthur Rothera, the first lecturer in Biochemistry, was also a qualified medical practitioner, enlisting in the Australian Army Medical Corps in 1915. Peter Hall was appointed in 1966 to the foundation Chair in Medical Biochemistry and was succeeded in 1973 by Gerhard Schreiber, who served in the position until 1997. There has been no professor of medical biochemistry since then. The number of medical graduates specialising in biochemistry has declined even more sharply in the Department of Biochemistry than in the Department of Physiology, where a few, notably Stephen Harrap, Graham Barrett and Mastura Monif, remain. Nonetheless, with its heavy emphasis on research into diseases such as cancer, neurodegenerative disorders, malaria, parasite infections and inflammatory autoimmune diseases, work in the Department of Biochemistry and Molecular Biology is intimately linked to that of researchers in other parts of the medical faculty.

The department has always been extremely successful in attracting funds from the NHMRC, as well as from industry bodies more closely aligned with its non-medical research. In 2013 alone, members of the department won grants amounting to over $8 million. Malcolm McConville was awarded almost $1.68 million in funds for a Project Grant and Research Fellowship to advance research into infectious diseases such as leishmaniasis and tuberculosis; Leann Tilley was awarded $693 458 to continue her team's investigation of the resistance of the malaria parasite *Plasmodium falciparum* to artemisinin-based antimalarial drugs; José Villadangos won a research fellowship for $727 610 that will allow further research into cells and molecules involved in antigen presentation, which has the potential to develop more effective vaccines and prevent autoimmunity, allergy and transplant rejection; and Stuart Ralph and Kathryn Holt won Career Development Fellowships that will allow them to further their respective work into parasitology and research investigating how bugs like bacteria cause infections and other diseases.[2]

Recent postgraduates have continued this record of achievement, both at the University of Melbourne and overseas.

Helena Safavi-Hemami (1979–) took her PhD entitled 'Understanding Synthesis, Folding and Modification of Conotoxins: Model Disulfide Rich Peptides' from the University of Melbourne in 2010. She took her MSc from the University of Cologne, during which time she spent nine months conducting her research at the Australian Institute of Marine Science in Townsville, and came to Melbourne by a more circuitous route than many of its postgraduates, having been awarded an International Postgraduate Research Scholarship to work with Tony Purcell and Bruce Livett on the biochemistry of small neuropeptide toxins (conotoxins) from predatory marine cone snails. Her thesis focused on the use of transcriptomics, proteomics and mass spectrometry to identify novel conotoxins and characterise enzymes responsible for their synthesis, modification and maturation.

In 2011 she was awarded a University of Melbourne Early Career Research Grant and continued her conotoxin research as a postdoctoral fellow in the Purcell laboratory. In 2012 she was one of two recent post-graduates from the department (the other was Cheryl Chia from the Gleeson laboratory) who were awarded ASBMB fellowships. Later that year she took a position as postdoctoral fellow at the University of Utah and in 2013 was awarded the prestigious Marie Curie Fellowship by the European Union. January 2014 saw the e-publication of an important finding in venom research, revealing new insights into the mechanism behind cone snail venom diversification, which will appear in *Molecular and Cellular Proteomics*.[3] In April 2014 she was promoted to research assistant professor and is continuing her conotoxin investigations in Professor Baldomero Olivera's group at the University of Utah.

Shayne Bellingham (1972–), who took his BSc in 1996, his BSc (Hons) in 1997 and his PhD (in the Department of Genetics) from Melbourne in 2005, is the inaugural Bellberry Indigenous Health Research Fellow. Bellberry Limited, which was formed in 2004, is a not-for-profit company that undertakes independent ethics reviews and monitoring of human research projects. The first fellowship, of $200 000, was awarded in 2012. Bellingham's doctoral thesis investigated copper homeostasis and the Alzheimer's disease amyloid precursor protein, and his current research builds on this work.[4] Specifically, he is interested in 'trying to capture genetic signatures and utilise this information from a diagnostic point of view, as early detection generally leads to better health outcomes'.[5] His investigations have begun with his own family, the Wotjobaluk people now living in the Horsham region in Victoria.

The Bellberry Fellowship is the latest in a long list of accolades that Bellingham's work has earned. In 2002 he won a Loxton Bequest Scholarship and awards from the Minerals Council of Australia and the Foundation for Young Australians, as well as an Alfred Nicholas Travel Scholarship. In 2006 he was awarded an ICAD Travel Scholarship from the American Alzheimer's Association, and in 2007 he received the Ann Miller Victoria Research Grant. In 2012 Bellingham and Andy Hill announced a discovery that could lead to a simple blood test for CJD and BSE, and which might also provide a non-invasive test to detect other neurodegenerative diseases, including Alzheimer's and Parkinson's.[6]

Shayne Bellingham is the inaugural Bellberry Indigenous Health Research Fellow and investigates copper homeostasis and the Alzheimer's disease amyloid precursor protein

The Wettenhall Establishment Award commemorates the inaugural director of the Bio21 Institute. Awarded annually to a postdoctoral researcher, it supports individuals with a strong research track record who are currently undertaking work aligned with one or more of the research themes of the Bio21 Institute. It is designed to assist with the transition from senior postdoctoral scientist to independent researcher. The winner of the 2014 award was Matthew Dixon (1980–), who took his BSc in 2002 and his PhD in 2008 from the University of Queensland. Having moved to La Trobe University and won both a C.J. Martin Research Fellowship and the Lorne Protein Structure and Function ECR Award in 2010, he came to Melbourne University in 2011, working on understanding the remodelling and cell biology processes that drive the unique architectural changes undertaken by the human malaria parasite *Plasmodium falciparum*. Also in 2011 he was awarded a Ramaciotti Foundation start-up grant and an ANZ Trustees Equipment Grant. He is an associate member of the ARC Centre of Excellence for Coherent X-ray Science (CXS).

The changes in *Plasmodium falciparum* being investigated are central to the parasite's ability to survive within the red blood cells and circulation of its human host. Dixon is particularly interested in the processes mediating cerebral and placental malaria and the mechanisms of sexual differentiation and transmission. Understanding these processes will help guide much needed new approaches aimed at disease eradication. These investigations led to a discovery reported in 2012 in the *Journal of Cell Science*.[7] The University of Melbourne's *Voice* noted:

Matthew Dixon won the Wettenhall Establishment Award and is investigating changes in *Plasmodium falciparum*

> One child dies from malaria every minute in Africa, and around the world, the malaria parasite kills more than 600 000 people each year, most of them children and pregnant women, while another 225 million people suffer illness as a result of malaria infections.
>
> In 1880, the malaria parasite was first observed in its banana form in the blood of a patient in Africa. For most of its life cycle, the parasite has a round shape within its human hosts, but has to take on a banana shape for sexual reproduction.
>
> Now, more than 130 years later, an Australian research team led by Dr Matthew Dixon and PhD student Megan Dearnley from the Department of Biochemistry and Molecular Biology, Bio21 Institute has revealed how malaria performs this shape-shifting.
>
> 'Using a 3D microscope technique, we reveal that malaria uses a scaffold of special proteins to form a banana shape before sexual reproduction', Dr Dixon says.

'As the malaria parasite can reproduce only in its "banana form", if we can target these scaffold proteins in a vaccine or drug, we may be able to stop it reproducing and prevent malaria transmission entirely.'

When in its banana shape, the malaria parasite is passed from a human host to a mosquito, where it reproduces in the mosquito gut. The study found that specific proteins form scaffolds, called microtubules, which lie underneath the parasite surface and elongate it into the sexual stage banana shape.

The work suggests that when the parasites are ready for sexual reproduction, they adopt the banana shape so that they can fit through the tiny sinusoidal slits in the spleen. This enables them to avoid the host's mechanical filtering and immune surveillance mechanisms and survive in the circulation long enough to be picked up by a mosquito and transmitted to the next victim.

The banana shape was revealed in greater detail than ever before by using high-end imaging techniques – 3D Structured Illumination Microscopy and Cryo Electron Microscopy – conducted with the ARC Centre of Excellence for Coherent X-Ray Science.[8]

Cheryl Chia (1984–) came to the University of Melbourne from Singapore with a Melbourne International Fee Remission Scholarship in 2003, took her BSc (Hons) in 2008 and was awarded a Melbourne International Research Scholarship for her PhD in the Gleeson laboratory.[9] A large number of awards, including an M.A. Bartlett grant, an International Student Travel Award from the American Society of Cell Biology, a 2001 Australian Society of Biochemistry and Molecular Biology Student Bursary Award and the David Walsh Student Prize (ComBio 2011), enabled Chia to present

Cheryl Chia took a position as postdoctoral fellow in the National Institute of Dental and Craniofacial Research, NIH, in the USA.

her research at interstate and international conferences. An ASBMB Fellowship in 2012 enabled her to attend the Lysosomes and Endocytosis Gordon Research Conference in Andover, New Hampshire.

After ten months as a postdoctoral research fellow in the department, where she examined the role of transmembrane domains in sorting at the early endosome, and explored the use of flow cytometry for high-throughput quantification of distinct trafficking pathways, Cheryl Chia took up a position as postdoctoral research fellow in the Chen WanJun Laboratory, National Institute of Dental and Craniofacial Research, National Institutes of Health in Bethesda, Maryland in 2013. The same year, she published a paper in *Traffic* on the intracellular itinerary of internalised beta-secretase BACE1 and its potential impact on amyloid precursor protein processing, work which has important implications for the aetiology of Alzheimer's disease.[10]

The research links at the University of Melbourne between Biochemistry and other departments and faculties have survived a remarkable amount of geographical relocation and dislocation, which shows little sign of abating. This has been exemplified by the battles fought by successive staff to achieve the best environment for their work, from Arthur Rothera, who raised funds from 1912 to 1914 for improvements to his laboratory and teaching space, to Victor Trikojus, who fought for the Russell Grimwade Building for a decade and a half before seeing the final stage opened. In 2005 the research staff moved to the David Penington Building, and in 2014 a Cell Signalling Centre was established in the Medical Building, joining the teaching-only staff and undergraduate laboratories there.

The enormous redevelopment of the south-western corner of the University of Melbourne campus places some of the activities of the Department of Microbiology and Immunology in the newly constructed Peter Doherty Institute. The new Victorian Comprehensive Cancer Centre, home to the Peter MacCallum Cancer Centre and new cancer research and clinical services for Melbourne Health and the University of Melbourne, which is expected to open in 2016, will provide new, easier links with researchers from many departments, notably from Biochemistry and Molecular Biology, who are working on cancer.

This book contains a brief account of the lives and work of a small number of the men and women who have contributed to the achievements of the Department of Biochemistry and Molecular Biology in just over three-quarters of a century. The department lists ambitious aims in its mission statement which involve both the emergence of new techniques and the establishment of new relationships. They include: to achieve biotechnology innovation through multidisciplinary research, and alliances with academia and industry; to attract outstanding scientists and technicians; and to establish core platform technology facilities accessible to diverse scientific and industry communities. It will strive to engage industry and nurture the commercialisation of discoveries, to support start-up companies through business incubation and the development of entrepreneurship skills, and to prepare research students and postdoctoral fellows for leadership in industry. It will also seek to translate research into educational and economic community benefits and provide a forum for community debate and dissemination of information on emerging bioscience and technology issues. Working to fulfil this mission will ensure that the Department of Biochemistry and Molecular Biology at the University of Melbourne remains one of the most important national and international bodies engaged in transforming biology.

Doctoral Degrees Awarded

Achen, Marc Gregory. A Study of Gene Expression in Streptococci. 1988.

Aitken, Cathy Jean. The Effect of the Level of Galactokinase Activity on Galactose Utilization in *Streptococcus thermophilus*. 1993.

Aldred, Angela Roslyn. Studies on the Synthesis and Secretion of Plasma Proteins. 1989.

Aldred, George Peter. Structure and Expression of the Ovine Renin and Kallikrein mRNAs. 1987.

Allen, Stacey. The Processing and Presentation of the Gastric H+/K+ ATPase for Maintaining Self-tolerance. 2011.

Alsop, Kathryn Jane. Molecular and Genotypic Analysis of BRCA1/2 Mutations in Ovarian Cancer. 2012.

Anders, Robin Fredric. Aspects of the Growth and Metabolism of Group N Streptococci. 1967.

Ang, Koon Yong. Desmond Autoimmune Gastritis Susceptibility Genes in Mice. 2007.

Apostolopoulos, Jim. The Effects of Thyroid Hormones on Apolipoprotein Gene Expression. 1989.

Appleby, Cyril Angus. The Cytochrome-linked Dehydrogenase Systems of Yeasts and Higher Plants. 1957.

Argyropoulos, Victor Peter. Studies on the Regulatory Protein TyrR from *Escherichia coli*. 1989.

Atcliffe, Benjamin William. The Structure and Interactions of Human Apolipoprotein C-I. 2006.

Atkinson, Sarah Catherine. Structure, Function and Inhibition of Dihydrodipicolinate Synthase from Bacteria and Plants. 2012.

Austin, Lawrence. The Mechanism of Action of Drugs Classed as Anticholinesterases. 1958.

Bachelard, Herman Stanton. Chemical and Biochemical Studies on Endemic Goitre in Australia. 1960

Bagnara, Aldo Sebastian. Regulation of Intracellular Nucleotide Concentrations. 1972.

Bailey, Michael Frederick. The Interaction of the *Escherichia coli* Regulatory Protein TyrR with DNA. 1999.

Baird, Cameron William. Studies on Plasma Insulin. 1959.

Baldwin, Graham Sherard. Studies on Chorismate Mutase-Prephenate Dehydratase from *E. coli* K12. 1974.

Baldwin, Graham Sherard. Growth Factors of the Gastrointestinal Tract: Collected Works 1981–2004. (DSc) 2004.

Baran, Katherine. A Functional Analysis of Perforin. 2008.

Barras, David Russell. A Study of the Pellicle Structure and the Metabolism of Paramylon in *Euglena gracilis*. 1968.

Baseggio, Nina. Regulation Studies on the *Escherichia coli* Gene aroG. 1992.

Beddoe, Travis. Surface Proteins of the Ribosome and Mitochondria Involved in Protein Import. 2003.

Belgi, Alessia. Chemical Synthesis and Structure-function Relationship Study of Insulin-like Peptide 5 (INSL5). 2012.

Bellair, John Terence. Studies on Histones. 1965.

Bergemann, Andrew David. Physical Studies in Mollicute Genetics. 1989.

Bidwell, Bradley N. The Role of the Immune System and Immunomodulatory Cytokines in Breast Cancer Metastasis. 2008.

Bilszta, Justin L.C. Heat Shock Proteins in Vascular Protection. 2001.

Birch, Helen Elizabeth. Studies on Acute Phase Protein Genes. 1986.

Bird, Megan Judith. Biochemical Characterisation of Src-mediated Phosphorylation of the PTEN Lipid Phosphatase. 2011.

Birt, Lindsay Michael. The Uptake and Metabolism of Amino Acids by a Plant Tissue Blood Plasma. 1957.

Bolden, Jessica Erin. Investigating the Molecular Basis of Histone Deacetylase Inhibitor-mediated Tumour-selective Cell Death. 2008.

Bornstein, Joseph. Studies on Diabetes Mellitus. (DSc) 1955.

Boyce, John Dallas. Molecular Biology of a Temperate Lactococcal Bacteriophage. 1994.

Bozinovski, Steven. Identification and Characterisation of a Novel Mechanism by which Akt is Regulated. 2000.

Brack, Charlotte Mary. Salivary Agglutination of *Streptococcus mutans* and Colonisation of Rat Tooth Enamel. 1986.

Bradbury, J.H. Studies on the Chemistry of Peptides, Polymers and Proteins. (DSc) 1967.

Branster, Marjorie Vivienne. Enzymic Studies in Relation to Tumour Development. 1958.

Brocklebank, Anne Margaret. A New Fluorescent Ligand for the Nucleoside Transporter. 1994.

Burgess, Antony Wilks. Chemistry of Glyceraldehyde-3-phosphate Dehydrogenase from *Jasus verrauxi*; Conformational Energy Calculations on Methylated Peptides. 1972.

Burgess, Benjamin Ross. Evolution and Regulation of Dihydrodipicolinate Synthase. 2012.

Bursac, Dejan. The Evolution of the Mitochondrial Protein Import Machinery. 2010.

Buyya, Smrithi. Structural and Functional Studies of Human Ap4A Hydrolase. 2008.

Byrt, Pauline Naomi. Studies on Zoospores of *Phytophthora cinnamomi*. 1980.

Bywater, Megan Julie. Disruption of RNA Polymerase I Transcription as a Strategy for the Treatment of Cancer. 2011.

Cahill, David M. Changes in Host Physiology Caused by *Phytophthora cinnamomi*. 1986.

Callander, Gabrielle Elizabeth. Viral Vector and Gene Silencing Strategies to Investigate Relaxin Family Peptides and Receptors of the Central Nervous System. 2010.

Cancilla, Michael Robert. Characterization of Genes Involved in Lactic Acid Production by *Lactococcus lactis*. 1995.

Cao, Yuan. BMP4 – a Metastasis Suppressor Gene in Breast Cancer. 2011.

Carbone, Francis Robert. Synthetic and Spectroscopic Studies of Repeating B-bend Models Based on Feather Keratin. 1984.

Ch'ng, Ai Lee. Biosynthesis of Amino Acids from Inorganic Nitrogen Sources in a Marine Green Alga. 1980.

Chan, Khai-Chew. Activation of Src Family Kinases by HIV-1 Nef and Its Inhibition by Csk Homologous Kinase. 2010.

Chan, Nickie C. Functional Analysis of Protein Import and Assembly Machineries of the Mitochondrial Outer Membrane. 2008.

Chang, Linus. Studies of the Structure and Function of Transthyretins. 1998.

Chappell, Roderick James. Pig Growth Hormone: Isolation and Radioimmunoassay in Blood Plasma. 1972.

Chatelier, Ronald Christopher. Transverse Organization of Biomembranes Studied by Fluorescence Quenching. 1984.

Chelladurai, Mohanathasan. Intermediate Forms of Secretory Proteins. 1980.

Chen, Xing. Physical Maps of Genomes of *Mycoplasma gallisepticum* PG31 and S6 with Some Functional Loci. 1991.

Chen, Yu-Yen. Isolation and Characterisation of CRE-binding Proteins in Rat Liver Nuclei. 1999.

Cheung, Nam Sang. Modulation of Adrenal Medullary Function. 1993.

Chia, Pei Zhi Cheryl. Dissecting Retrograde Transport Pathways in Development and Disease. 2011.

Chong, Yuh Ping. C-terminal Src Kinase Homologous Kinase (CHK): a Potential Tumour Suppressor Inhibiting Oncogenic Src-family Kinase Activity. 2006.

Christopherson, Richard Ian. Interrelationships of Pyrimidine Biosynthesis in *Escherichia coli* K12. 1976.

Civciristov, Srgjan. Structure and Function of the TimA and TimB Transporter Proteins in the Cytoplasmic Membrane of the α-proteobacterium *Caulobacter crescentus*. 2012.

Claasz, Antonia Alexandra Georgina Cecilia. Properties of the Human Relaxin Receptor. 2003.

Clarke, Adrienne Elizabeth. Studies on β-Glucans and β-Glucoside Hydrolases. 1962.

Clements, Craig Steven. Expression and Heparin Binding of Recombinant Rat Hepatic Lipase. 2000.

Cocks, Benjamin Graeme. Metabolism and Genetics of Mycoplasmas, Particularly *Ureaplasma urealyticum*. 1988.

Cole, Adam Robert. Development of a Protein Database of Normal, Hyperplastic and Malignant Colonic Tissue. 2004.

Cole, Timothy James. The Structure and Regulation of Major Acute Phase α_1 Protein (Thiostatin) from Rat Liver. 1987.

Coleman, Bradley Maurice. Defining the Molecular and Structural Components of Exosomes that Modulate the Intercellular Spread of Prions. 2013.

Cooper, Helen Mary. The Structural Basis for the Immunological Distinction between 'Self' and 'Non-self' in Myoglobins. 1985.

Corlett, Alicia R.I. Defining the Interaction between the ER Resident HSP70 Chaperone Kar2p and the Proximal Sensor for the UPR Ire1p. 2004.

Cornish, Edwina Cecily. The TyrR Gene of *Escherichia coli* K-12. 1984.

Coventry, Michael John. Aspects of the Metabolism of *S. faecium* and *S. cremoris*. 1983.

Cowan, Sandra Wendy. Crystallographic Studies of Platelet Factor 4. 1988.

Crewther, W.G. Physical and Chemical Studies on the Structure of Wool and Other Research. (DSc) 1969.

Cristiano, Briony Elizabeth. Characterisation of the Expression and Activity of Akt3 in Ovarian Cancer. 2004.

Curtain, C.C. Studies on Abnormal Serum Globulins. (DSc) 1965.

D'Auvergne, Edward James. Protein Dynamics: a Study of the Model-free Analysis of NMR Relaxation Data. 2006.

Dagley, Michael James. Protein Translocation in the Remnant Mitochondria of *Giardia intestinalis*. 2008.

Dalgarno, Lynn. Respiratory Metabolism and Processes of Uptake in a Plant Tissue. 1962.

Davidson, Barrie Ernest. The Interaction of Glutathione with Cystamine and Protein Disulphide Bonds. 1965.

Davies, William David. Studies on the Structure and Expression of aroG, the Gene Coding for 3-deoxy-D-ara.b\noheptulosonate-7-phosphate synthetase (phe) in *Escherichia coli*. 1983.

Davis, Amanda Jane. A Study of the Effect of Phosponates on the Metabolism of *Fusarium spp*. with Particular Reference to *Fusarium oxysporum f.sp. cubense*. 1995.

Davoren, P. R. The Action of Protein Hormones on the Metabolism of the Isolated Mammalian Liver with Particular Reference to the Action of Glucagon. 1957.

Davuluri, Siva Prasad. A Re-appraisal of the Function of AMP-aminohydrolase and of the Function and Synthesis of Phosphagens in Skeletal Muscle. 1982.

Dawson, Desmond John. Immunochemical Studies of Tetanus Toxin. 1967.

De Jong, Felice Aleida. Plasma Protein Gene Expression during Inflammation, Fasting and Protein Depletion. 1987.

De Pury, Guillaume George. Some Studies on the Functions of Essential Fatty Acids and Phospholipids. 1965.

Deliyannis, Paris Plutarch. Analysis of the Interactions of Human Apolipoprotein A-l using Sedimentation Equilibrium and Monoclonal Antibodies. 1992.

Demchyshyn, Vasyl. Characterization of Novel Proteins in the Mitochondrial Outer Membrane. 2004.

Denborough, M.A. Studies on Malignant Hyperpyrexia and on Glycoproteins in Man. (DSc) 1977.

Derby, Merran Clare. The Function of GRIP Domain Golgins in Membrane Trafficking and Golgi Organization. 2006.

Dickson, Phillip Wesley. Rat Transthyretin: Biosynthesis and Metabolism. 1985.

Diepenhorst, Natalie. A Novel Approach for Investigating GPCR/ligand Interactions: Analysis of the Secondary Binding Site within RXFP1. 2013.

Diesch, Jeannine. Characterisation of Pro-Invasive Transcriptional Programs Regulated by the Fra-1 Transcription Factor. 2012.

Dixon, Matthew P. Characterisation of the Ligand Response Domain of the Transcriptional Regulator TyrR from *Escherichia coli*. 2008.

Dobric, Nada. Characterization of *Lactococcus lactis* Genes Involved in Methionine and Cysteine Metabolism. 2002.

Dopheide, Theodorus Antonius Aloisius. Studies on the Specificity of Adrenal Acid Proteinase: Investigation of the Action of Thyroid Acid Proteinase, with Particular Reference to Thyroglobulin. 1965.

Dorow, Donna Sutton. Conformational and Flexibility Factors in the Immunoreactivity of Beef Myoglobin and Its Peptides. 1986.

Dow, Lukas E. The Role of Scribble in Polarity, Migration and Tumorgenesis. 2006.

Dreher, Theo Wolfgang. Studies on the Wound Response of the Coenocytic Green Alga *Caulerpa simpliciuscula*. 1980.

Duan, Wei. Transthyretin: Structure, Gene Expression and Evolution. 1991.

Dunkley, Peter Robert. Structure and Function of the Basic Proteins of Rat Brain Myelin. 1974.

East, Iain James. The Structural Basis of Antigenicity in Myoglobins. 1980.

Edwards, Kaylene Joy. Biosynthesis of Albumin in Rat Liver. 1978.

Ehlgen, Florian. The Plasmodium Food Vacuole: Protein Targeting of Transmembrane Proteins. 2011.

Elsum, Imogen. Regulation of Oncogenic Ras-MAPK Signalling by the Polarity Protein Scribble. 2011.

Emanuelle, Shane. The Regulation of SnRK1 from *Arabidopsis thaliana*. 2013.

Etemadmoghadam, Dariush. Genomic Variation in Primary Chemoresistant ovarian Carcinomas. 2008.

Evans, David John. Synthetic Peptide Models for Chain Reversal Conformations in Proteins. 1980.

Farrer, Keith Thomas Henry. The Determination and Thermal Destruction of Vitamin B1: a Collection of Papers. (DSc) 1954.

Finch, Lloyd Ross. The Active Transport of Amino Acids by Intestinal Tissue of the Rat. 1959.

Fincher, Geoffrey Bruce. Studies on Polysaccharides in Wheat Endosperm. 1973.

Francis, Christopher John. Insight into the Early Stages of Protein Folding from Structural Models of the Unfolded State. 2007.

Franic, Teo Vicko. Epithelial Cell Dynamics in Mice Deficient in Gastrin and the Gastric H/K-ATPase. 2004.

Frater, Robin. Protein Thiols and Disulphides in the Rheology of Flour Systems. 1962.

Fullerton, Paul Douglas. Protein Synthesis and the Effect of Amino Acid Analogs in *Pseudomonas aeruginosa*. 1970.

Fung, Wai-Ping. Regulation of Eukaryotic Gene Expression and Protein Synthesis. 1987.

Gabriel, Kip. A Study of the Dynamic Interactions between the Components of the Pre-protein Translocase of the Outer Mitochondrial Membrane. 2004.

Gajewski, Joanna. Biochemical Basis of Inactivation of the Tumour Suppressor PTEN by Frequently Encountered Tumour Associated Mutations. 2003.

Ganella, Despina Electra. A Viral Based Strategy to Modulate the Relaxin-3 Receptor in the Rat Brain. 2013.

Gentle, Ian Edward. Evolutionary Origins and Mechanisms of Mitochondrial Outer Membrane Assembly. 2006.

Georgakopoulos, Thanae. The Regulation of Monocyte/macrophage Responses by Colony Stimulating Factors and Inflammatory Mediators during Cellular Interactions. 2005.

George, Joshy. Integrated Analysis of Distinct Molecular Profiles of Ovarian Cancer. 2013.

Gething, Mary-Jane H. Structural and Physicochemical Studies on Chorismate Mutase-prephenate Dehydratase from *E. coli* K12. 1973.

Gibbs, Gerard Mark. Structure and Function of Piscicolin 126. 2002.

Gibson, Frank William Ernest. Studies on the Synthesis of Aromatic Compounds by Bacteria. (DSc) 1963.

Gilan, Omer. Unravelling the Nature of Fra-1 Transcription Factor Complexes in Invasive Tumour Cells. 2012.

Gillespie, J.M. Collected Papers – Studies on the Proteins of Wool Keratin. (DSc) 1963.

Gireesh, Tulasiram. Conjugal Transfer and Expression of Genes Encoding Nisin Production in *Lactococcus lactis*. 1992.

Gloriani-Barzaga, Nina. Studies on Antigenic Diversity in the Malaria S-antigen. 1987.

Gorasia, Dhana Govind. Molecular Events that Render Pancreatic Beta Cells Targets of T Lymphocytes. 2012.

Gottschalk, Alfred. Collected Papers for DSc 1949. 1950.

Gould, Michael Kun. Biochemical Adaptation in Response to Physical Exertion. 1961.

Greening, David. Peptidomics: Isolation and Characterisation of Endogenous Peptides. 2008.

Grubb, Darren John. Structure/function Analysis of the Human Carbonic Anhydrase VI Isozyme & Analysis of the Human Carbonic Anhydrase VI Gene. 2000.

Grusovin, Julian. Munc18c and Insulin Action. 2006.

Gunawardana, Dilantha. Characterisation of the mRNA Decapping Enzyme and Related Nudix Hydrolases from *Arabidopsis thaliana*. 2007.

Gunn, Natalie Johanna. New Insight into the Role of the SH_2 Domain in the Regulation of Src Family Kinases and Their Endogenous Inhibitor, Csk-Homologous Kinase. 2012.

Gunn, Priscilla Anne. Understanding Regulated Transport and Specialised Endo-membrane Compartments in Parietal Cells, the Acid Secreting Cells of the Stomach. 2010.

Gunnersen, Jennifer M. Characterization of Human and Rat Relaxin Gene Expression. 1994.

Gunzburg, Menachem Joseph. The Interaction of Apolipoprotein E with Amyloid Fibrils. 2007.

Haigh, Elizabeth Anne. Protein–lipid Interactions in Biomembranes as Studied with a Set of Fluorescent Fatty Acid Probes. 1980.

Han, Sen. Revealing the Cryptic Bityrosine Motifs in the Mammalian Prion Protein. 2010.

Harford, M.N. Transformation and Chromosome Replication in *Bacillus subtilis*. 1966.

Harrison, Christopher F. The Hydrophobic Region of PRP and Its Role in Prion Disease. 2010.

Hatters, Danny Martin. Characterisation of Human Apolipoprotein c-II Amyloid Fibrils. 2002.

Hawthorne, Donn Bede. Studies on the Structure and Function of Polysaccharides from the Green Alga *Caulerpa simpliciuscula*. 1980.

Henderson, Robert William. Electromotive Force and Related Physico-chemical Studies on Metallo-porphyrin Systems. 1960.

Heng, Joanne Soo Ping. Hexosamine-dependent Growth and Virulence in *Leishmania major*. 2010.

Hennebry, Sarah Catherine. The Evolution of the Structure and Function of Transthyretin-like Protein. 2007.

Henning, Susan June. Volatile Fatty Acids and Their Metabolism in Fermentative Organs of Some Herbivores. 1971.

Hickey, Malcolm Wayne. Aspects of the Metabolism of Homofermentative Lactobacilli. 1982.

Hildebrand, Michael Stephen. Molecular Mechanisms of Auditory Function and Stem Cell Mediated Regeneration of the Inner Ear. 2006.

Hill, John Stephen. Nature of the Membrane Defect in Hereditary Spherocytosis of Man. 1983.

Hillier, Alan James. Aspects of the Metabolism of Carbon Dioxide by *Streptococcus lactis*. 1976.

Hinde, Richard Wesley. Boron Nutrition in Plants with Special Relation to Nucleic Acid Metabolism. 1965.

Hird, Francis John Raymond. Collected Papers in Biochemistry. (DSc) 1962.

Hitchcock, Michael James Martin. Aspartokinases of *Bacillus brevis* ATCC 10068. 1976.

Hogan, Thea Valmai. Tolerance and Immunity to the Gastric H+/K+ ATPase. 2008.

Hogg, David McColl. The Antibacterial Action of Lactoperoxidase. 1968.

Holloway, Andrew James. Genetic and Biochemical Characterisation of the Murine Siah Genes. 1997.

Holton, Timothy Albert. Petal-specific Gene Expression in Petunia Hybrid. 1992.

Hoogenraad, Nicholas Johannes. Studies on the Contribution of Rumen Bacteria to the Nutritional Requirements of Sheep. 1969.

Hopkins, Emma June. The NMR Solution Structure of the RXFP1 Receptor LDLa Module and Identification of Residues Involved in Receptor Signaling. 2007.

Hor, Lilian. Structural and Functional Characterisation of Diaminopimelate Epimerase from *Escherichia coli*. 2012.

Hossain, Mohammed Iqbal. Investigating the Role of c-Src in Excitotoxic Neuronal Death. 2012.

Howard, Kristy Ann. Biochemical and Genetic Studies of D Hordein in Barley and Its Relationship to Malting Quality. 1995.

Howard, Russell John. Products of Photosynthesis in the Marine Green Alga *Caulerpa simpliciuscula*: Studies on the Intact Plant and with Isolated Chloroplasts. 1976.

Howells, Antony John. A Study of Some Aspects Related to Protein Synthesis during Metamorphosis and Development in the Sheep Blowfly *Lucilia cuprina*. 1965.

Howlett, Geoffrey John. The Sedimentation Equilibrium of Chemically Interacting Systems with Special Reference to Studies on Lysozyme. 1972.

Hudson, Bryan. Studies on Adrenal and Pituitary Hormones: the Nature of Hormone Induced Eosinopenia; Observations on the Pigmentary Disturbance in Addison's Disease. 1957.

Hudson, Graham Stanley. Studies on Chorismate Mutase/prephenate Dehydrogenase from *Escherichia coli* K12. 1982.

Hulett, Joanne Maree. Evolutionary Conservation and Aberration Reveal Functionally Important Domains in the Mitochondrial Protein Import Receptor Tom20. 2007.

Hung, Sandy Shen-Chi. Regulation of Ribosomal Gene Transcription during Cancer: Is Up-regulation of 45S rRNA Synthesis Sufficient to Drive Growth and Proliferation? 2009.

Hurrell, John Gordon Rowan. Immunological and Structural Studies on Some Plant and Animal Globins. 1977.

Hurt, Pauline Ruth. Binding and Reactive Sites in Serum Albumins. 1972.

Husain, Athar. Protein Interactions in the Erythrocyte Cytoskeleton. 1984.

Ia, Kim Kui. The Regulation and Protein Substrate Selectivity of CSK-homologous Kinase (CHK): an Endogenous Inhibitor of Oncogenic SRC-family Tyrosene Kinases (SFKS). 2009.

Ilgoutz, Steven Christopher. Characterisation of the Conglutin [Gamma] Gene from *Lupinus angustifolius*. 1996.

Iliades, Peter. The Design and Engineering of Antibody-like Molecules and Their Molecular Interactions. 1997.

Irtegün, Sevgi. Tracking Open and Closed Conformations of c-Srs in the Cell Using Fluorescent Biosensors. 2012.

Irving, Helen Ruth. Mechanisms Involved in the Differentiation of Phytophthora Zoospores. 1984.

Iverson, W.G. Bacteriophages and Bacteriocins of *Streptococcus bovis*. 1976.

Jablonski, Paula. The Distribution of Proteolytic Enzymes in the Thyroid Gland. 1967.

Jackson, Warwick John Hamilton. A Study of the Binding of Ligands to Self-Interacting Acceptor Molecules with Special Reference to Studies on α-Chymotrypsin. 1969.

Jago, George Richard. Nutritional Aspects of the Growth of Lactic Acid Streptococci. 1957.

James, Patrick F. Lipid–lipid and Apolipoprotein–lipid Interactions in the Gas and Solution Phases. 2008.

Jamieson, Gary Peter. Transport and Metabolism of Cytosine Arabinoside in Human Tumour Cells. 1994.

Jastrzebski, Katarzyna. Role of S6K Signaling in the Regulation of rRNA Gene Transcription, Ribosome Biogenesis and Cellular Growth. 2009.

Johnson, Elizabeth Dyer. Synthesis of Storage Globulins in *Lupinus augustifolius*. 1986.

Johnson, Timothy M. Regulation of Src-family Protein Kinases by Phosphorylation and Protein–protein Interactions. 2000.

Jokubaitis, Venta Joanna. The Role of Angiotensin Converting Enzyme and the Renin-angiotensin System in the Regulation of Haemopoiesis. 2006.

Jones, Ian Keith. Disulphide Peptides in Flour. 1968.

Kamaruddin, Aizuddin Mohd. Novel Peptide-based Approaches to Study the Activity and Substrate Specificity of Protein Kinases. 2013.

Keizer, David Wicher. The Role of Elicitins in the Defense Responses of Plants. 1998.

Ketaren, Natalia Elena. Investigating the Structure, Function and Regulation of Dihydrodipicolinate Synthase from *Streptococcus pneumoniae*. 2011.

Khalil, Zeinab. Role of Tachykinins in Catecholamine Secretion. 1987.

Klonis, Nectarios. A Spectral Analysis of an Antigen–antibody Interaction at an Interface. 1998.

Knesel, Jaromir A. Studies on the Role of Gastrin as a Growth Factor for the Gastrointestinal Tract. 1986.

Koay, Ann. Structure and Function of the ß-CBMS of AMPK. 2012.

Kolivas, Sotirios. Analysis of the cDNA Amino Acid and Promoter Sequences of Genes Coding for the Sulphur-rich Storage Proteins in the Seeds of *Lupinus angustifolius*. 1992.

Kong, Roy Chze Khai. Studies on the Role of the Low Density Lipoprotein Class A (LDLa) Module in the Activation of the Relaxin Family Peptide Receptors 1 and 2. 2012.

Kotsonis, Steven Efstratios. Detection and Characterization of Bacteriophages Infecting *Lactococcus lactis*. 1999.

Krieger, Vera Isabel. Published Articles and Those Accepted for Publication Submitted for Examination for the Degree of Doctor of Science on March 7th 1938. 1938.

Kwok, Terry. The Domain Structure of the Regulatory Protein TyrR from *Escherichia coli* K-12. 1998.

Lambert, Paul Francis. The Chemical Synthesis of Primate Relaxins. 1988.

Lanigan, Mark Damien. Structure and Design of Analogues of the Potassium Channel Blocker, ShK Toxin. 2000.

Lau, Kim. A Proteomic Approach to Investigate the Pathogenesis of Rheumatoid Arthritis. 2012.

Laurie, Karen Louise. Dissection and Manipulation of Autoreactive T Cell Responses in Autoimmune Gastritis. 2002.

Layden, Selena S. Isolation and Characterisation of Porcine Prorelaxin Processing Enzymes. 1996.

Lee, Rachel Szu-Hui. The Characterisation of AKT Isoform-specific Signalling. 2010.

Lees, Gordon John. The Role of Acetaldehyde in the Metabolism of Group N Streptococci. 1969.

Lekgabe, Edna Dieudonne. Cardiovascular Disease: Therapeutic Intervention with Relaxin. 2007.

Lennie, Robert William. Biochemical Aspects of the Development of the Blowfly, *Lucilia cuprina*. 1967.

Lennox, Francis Gordon. Publications with References. (DSc) 1941.

Li, Qiu-Tian. Lipid Dynamics in Emulsions. 1990.

Lieu, Zi Zhao. Regulation of Membrane Transport Pathways by GRIP Domain Golgins. 2008.

Lim, Jet Phey. Sorting Nexin 1 and Sorting Nexin 5 in Macropinocytosis and Membranc Trafficking. 2010.

Lim, Justin Wee Eng. Proteomic Insights into Secretome of Colorectal Cancer Cells. 2010.

Lin, Jane I Chun. Novel Roles for Ribosomal Protein Mutants in Regulating Growth and Proliferation in *Drosophila melanogaster*. 2010.

Lin, Liang-Ju (Jason). Structure and Function of a Lipid and Protease Activated Protein Kinase (PRK1). 2004.

Liu, Junjun. Cloning, Expression and Characterization of a Novel Rat Protein Kinase, prk2. 2002.

Liu, Su-Ting. Synthesis and Evaluation of Non-viral Delivery Vectors for Antisense Oligonucleotides. 2003.

Longano, Sandro. Sedimentation Equilibrium Analysis of the Interaction of Plasma Proteins and Lipoproteins. 1992.

Loughlin, Richard Edward. Purification and Properties of Thyroid Cysteinyltyrosinase. 1961.

Lueder, Franziska Birte. On the Proteins Involved in Mitochondrial Morphology and Membrane Assembly. 2010.

Luke, Michael Ross. The Mammalian Grip Domain Proteins: a Family of Golgins Localised to Subdomains of the Trans-Golgi network. 2006.

Ma, Sherie. Relaxin Peptide and Receptor Systems in the Rat Central Nervous System: Neuroanatomical, Behavioural and Gene Regulation Studies. 2005.

Macasev, Diana. Targeting of Tail-anchored Proteins to Mitochondria. 2003.

Macphee, Cait. The Specificity of Lipid–protein Interactions. 1999.

Mahanivong, Chitladda. Molecular Biology of a Temperate Lactococcal Bacteriophage. 2002.

Maksel, Danuta Maria. Cloning, Expression and Mutational Analysis of Recombinant Diadenosine Tetraphosphate Hydrolase from *Lupinus angustifolius*. 2001.

Malby, Robyn Louise. Protein Engineering and X-ray Crystal Structure of an Antibody-antigen Complex. 1994.

Mares, Daryl John. Chemistry and Organization of Endosperm Cell Walls. 1972.

Maroudas, Peter. Characterisation of the TyrR Protein from *Escherichia coli*: ATPase Activity and Interactions with DNA. 1999.

Martin, Roger F. Estimation of Deoxyribonucleic Acid. 1970.

Martin, Thomas John. Collected Papers in Biochemical Endocrinology: the Action and Metabolism of Calcitonin and Parathyroidhormone; the

Secretion of Hormones by Endocrine and Non-endocrine Tumour Cells; and the Effects of Hormones on Receptromediated Processes in Tumour Cells. 1978.

Martyn, John Charles. Cloning and Sequencing of the Genome of Feline Panleukepenia Virus (Feline Parvovirus). 1989.

Mathias, Rommel Andre. Proteomic Analysis of Extracellular Modulators of Epithelial-mesenchymal Transition. 2009.

Matkovich, Scot J. Inositol Phosphate Generation in the Heart: Mechanisms and Functional Relevance. 2000.

McKenzie, Geoffey Hamilton. Physicochemical Studies on Concanavalin-A, the Phytohaemagglutinin from Jack Bean. 1973.

McLeod, Margaret Frances. Studies on Ribonucleases. 1965.

McNaughton, Jean Wise. A Study of the Energy Expenditure and Food Intake of a Group of Adolescents. 1965.

McQuillan, Mary T. Biochemical Studies on the Thyroid Gland. 1952.

Menzies, Catherine Elizabeth. Studies on the Thyroid Gland. 1963.

Mertens, Haydyn David Thomas. Elucidation of the High Resolution Solution Structure of a Trypanosomatid Fyve Domain. 2004.

Milland, Julie. Regulation of Gene Expression in Regenerating Liver. 1991.

Mills, Ryan Daryl. Biochemical and Biophysical Analysis of Enzymatic Domains of the Parkinson's Disease-causative Protein Leucine-rich Repeat Kinase (LRRK2). 2011.

Misheva, Mariya. Characterisation of cJun N-terminal kinase 1 (JNK1) – Subcellular Localization, Post Translational Modifications and Protein Complexes under Control and Stress Conditions. 2013.

Mitchell, Alana. The Ribonucleotides of *Mycoplasma mycoides*. 1976.

Mitchell, Claire Louise. Effects of SRY Mutations on Its DNA-binding Activities and Nuclear Import. 2002.

Mitchelson, Keith Richard. Investigations on Malignant Hyperpyrexia. 1974.

Mok, Yee-Foong. Structure and Self-assembly of Human Apolipoprotein C-II Amyloid Fibrils. 2007.

Monk, Julie Alexandra. Thyroid Hormone Homeostasis in Developing and Adult Transthyretin Null Mice. 2006.

Monks, Stephen Andrew. Structure and Structure–function Relationships of Sea Anemone Cardio-stimulants. 1996.

Montalto, Joseph. C-19 Steroid Conjugates in Female Clinical Hyperandrogenism. 1990.

Mookerjee, Ishanee. Investigating the Role of Endogenous Relaxin in Experimental Models of Fibrosis. 2007.

Moore, Anne Elizabeth. A Glucan Hydrolase from *Nicotiana glutinosa*. 1970.

Moore, Christopher Hastings. Physico-chemical Studies on the Haemocyanin from *Jasus lalandii*. 1967.

Morgan, Francis Joseph. Human Chorionic Gonadotropin: a Study on the Structure and Properties of the Hormone and Its Subunits. (MD) 1975.

Moritz, Robert Lorenz. Structure–function Studies of Interleukin-6 and Its Receptor Complex. 2005.

Morrison, John Roderick. Determination of the Structural Domain of Apolipoprotein A1 Recognised by the High Density Lipoprotein Receptors. 1990.

Morton, Craig James. 1H-NMR Studies of Recombinant Murine Interleukin-6 and Related Molecules. 1994.

Morton, Donald James. The Oxidation of Amino Acids by Mitochondria. 1960.

Morton, Trevor Charles. Studies on the Prosthetic Group of Cytochrome *c*. 1970.

Mullin, Kylie Anne. Synthesis and Intracellular Transport of Glyco-sylphosphatidylinositol Anchored Molecules in *Leishmania Mexicana*. 2002.

Mustafa, Tomris. Characterisation of the Human AT4 Receptor. 2002.

Naderer, Thomas. Investigations into Cell Surface Intracellular Virulence Factors in *Leishmania Mexicana*. 2004.

Nagashima, Mariko. Studies on the Synthesis and Secretion of Glycoproteins. 1981.

Narasaiah, Gavini. Investigations on the Regulation of the *Escherichia coli* gene pheA. 1988.

Neale, Geoffrey Arnold Martin. Deoxyribonucleotide and One-carbon Metabolism in *M. mycoides subsp. mycoides*. 1984.

Ng, Milica. ^{13}C-based Metabolic Flux Modelling in Lieshmania Parasites. 2012.

Ng, Wai Hung William. Leishmania Metabolism and Virulence. 2013.

Ngyuen, Nhung Van. Gastric Acid Secretion and Cellular Homeostasis in the Gastric Mucosa. 2009.

Nicola, Nicos Anthony. Structural Studies on Leghemoglobins and Myoglobin. 1975.

Nisbet, Rebecca Maree. Defining a Role for the C-terminus of the Prion Protein in Prion Propagation. 2009.

O'Dea, Kerin. Physico-chemical Studies on the Lactate Dehydrogenase of *Streptococcus cremoris*. 1971.

O'Donnell, I.J. Chemical and Physicochemical Studies on Soluble Proteins from Keratins. (DSc) 1975.

Ojaimi Caroline. Molecular Genetic Analysis of Bacteria of the Genus *Borrelia*. 1998.

Pal, Bhupinder. Degradation of the Hac1p Transcription Factor Regulates the Unfolded Protein Response in *Saccharomyces cerevisiae*. 2008.

Parker, Lorien Jane. Crystallographic Studies of the Recognition of Anti-cancer Compounds by Glutathione Transferase Pi. 2009.

Paterson, Yvonne. The Effect of Side Chain Complexity on the Dimensions of Polypeptide Chains. 1978.

Peet, Daniel J. Protein-bound Fatty Acids in Mammalian Hair Fibres. 1994.

Peisley, Alys. MMP-2 Collagen Interaction and Exosite Inhibitin. 2005.

Peng, Chun-Ming Benjamin. Structure and Expression of a Liver Protease-activated Protein Kinase. 2000.

Perry, Andrew J. The Structure of Outer Mitochondrial Protein Import Receptors. 2007.

Perugini, Matthew Anthony. Characterization of Human Apolipoprotein E3 and E4 Isoforms. 2001.

Pham, Chi Le Lan. The Formation and Structure of Human Apolipoprotein C-II Amyloid Fibrils. 2006.

Phelan, David Ronald. The Role of Mixed Lineage Kinase 2 in Signal Transduction. 2003.

Phillips, Darren P. Responses of Susceptible and Resistant Avocado Cultivars to Infection by *Phytophthora cinnamomi*. 1989.

Phillips, John Williams. A Comparative Study of Ketogenesis and Gluconeogenesis. 1976.

Pierscionek, Barbara Krystyna. The Effects of Development and Ageing on the Structure and Function of the Crystalline Lens. 1988.

Ping, Fu. Expression and Characterisation of Rat Parathyroid Hormone-Related Protein (Rpthrp) in Yeast and Mammalian Systems. 2000.

Polidano, Megan Ann. Characterisation of the Neuronal Nicotinic Acetylcholine Receptor in Bovine Chromaffin Cells. 2002.

Poon, Carole. Investigation of the Role of Src Kinases in Cooperative Tumourigenesis with RasACT in *Drosophila*. 2008.

Powell, Ian Brinley. Studies on Lytic Bacteriophages of Lactic Streptococci. 1988.

Prapunpoj, Porntip. Evolution of Transthyretin (Gene Structure, Mechanism of Evolution of the N-terminal Region and Synthesis in *Pichia pastoris*). 1999.

Pulakat, Lakshmi Devi. Molecular Biology of Temperate Streptococcal Phage. 1988.

Qiu, Wen. Cellular, Molecular and Genetic Characterisation of Carcinoma-associated Fibroblasts from Breast and Ovarian Cancers. 2010.

Quinsey, Noelene S. Structural and Functional Analysis of Tyrosine Hydroxylase. 1997.

Radford, Ian R. DNA Synthesis Inhibition and Cell Death. 1980.

Rao, Zi-He. Refinement of Black Swan Goose-type Lysozyme. 1989.

Rasmussen, Richele Kim. Biological and Structural Characterisation of the Mixed Lineage Kinase 2 – Src Homology 3 Domain. 1998.

Read, Richard Stanley Docker. Studies on Biosynthetic Activities in Cell Nuclei. 1968.

Redvers, Richard Paul. Functional Characterisation of Candidate Keratinocyte Stem Cells. 2005.

Reid, Sonali Pearl. Investigation of the Role of Phosphonate in the Defence Response of Plants Using Defence Response Genes as an Index. 1997.

Reilly, Charles B. Targeted Discovery of T Cell Epitopes: High Energy, Stable Isotope Differentiated Precursor Ion Scanning. 2012.

Rellos, Peter. Regulation of the pheA Operon in *Escherichia coli*. 1982.

Reynolds, Eric Charles. Thymidine Sensitivity of Murine Myeloma and Lymphoma Cells. 1977.

Richardson, Samantha Jane. Evolution of Extracellular Thyroxine-distributor Proteins in Vertebrates. 1995.

Rimmer, Kieran. Structural Perspectives on Mitochondrial Import across the Outer Membrane. 2010.

Roche, Peter John. Structure and Expression of the Ovine Corticotropin-releasing Factor Gene. 1986.

Rosenberg, H. Studies on the Metabolism of Certain Phosphorus-containing Compounds. (DSc) 1969.

Ross, Ellen Margaret. Autoantigen-specific Regulatory T Cells in Autoimmunity. 2012.

Rudge, Geordie. Regulation of Anti-tumour Immune Responses In Vivo by CD4+ CD25+ Regulatory t Cells. 2006.

Ryan, Timothy Marc. The Effect of Lipids on Self-association and Fibril Formation by Apolipoprotein C-II. 2009.

Saad, Mirette Makram. Biological Characterisation of Human Phosphatidylinositide 3-Kinase Mutations. 2010.

Safavi-Hemami, Helena. Understanding Synthesis, Folding and Modification of Conotoxins: Model Disulfide Rich Peptides. 2010.

Sandall, David William. Isolating and Characterising Conotoxins from Australian *Conus spp.* 2006.

Saunders, Eleanor Clare. Investigations into the Metabolism of the *Leishmania* Parasite Using a Metabolomic Approach. 2008.

Scanlon, Denis Bernard. Part 1: Solid Phase Synthesis of Oligonucleotides by the Phosphotriester Method: Part II: Enzyme Substrate Peptide: a Model Peptide for the Elucidation of Side-reactions in Solid Phase Peptide Chemistry. 1984.

Scott, Daniel James. Characterizing the Extracellular Domains of the Relaxin and INSL3 Receptors, LGR7 and LGR8. 2006.

Shamgar, Felix Alexander. Phosphoglyceride Metabolism in Skeletal Muscle in Control Rats and in Rats Deficient in Essential Fatty Acids. 1974.

Shen, Peiyan. Regulation of C Apolipoprotein Gene Expression and Metabolism. 1990.

Shields, Benjamin James. Investigation of the Role of Heparin Binding Domains of Respiratory Syncytial Virus (RSV) G Protein. 2003.

Sidek, Hasidah Mohd. Tyrosine Phosphorylation and Signal Transduction in the Adrenal Medulla. 1994.

Sidhu, Gurcharn Singh. Amino Acid Metabolism with Special Reference to Uptake by Body Tissues. 1955.

Sim, Chou Hung. Investigating the Role of PINK1 in Parkinson's Disease. 2010.

Simpson, Richard John. Purification and Properties of 3-Deoxy-Arabino-heplulosonic Acid-7-Phosphate Synthetase (phe) from *Escherichia coli* K-12. 1973.

Sims, Neil Raymond. – 2', 3'-Cyclic Nucleotide 3'-Phosphodiesterase. 1977.

Sin, Iris L. Control of Nucleotide Biosynthesis in *Mycoplasma mycoides* and *Escherichia coli*. 1973.

Sinclair, Andrew James. The Role of the Essential Fatty Acids in the Lipid Metabolism of Rats. 1968.

Small, David Henry. In Vivo Methylation of Myelin Basic Protein in Relation to Neurological Disease. 1981.

Smith, David Keith. Structural Studies of Leukaemia Inhibitory Factor (LIF) and Its Receptors. 1996.

Smith, Geoffrey David. The Effects of Self-interaction of Ligand in Binding Processes with Special Reference to Studies on Thyroid Acid Proteinase. 1968.

Smith, John Alexander. Synthesis, Structure and Immunochemistry of Peptide Sequences in Globins and Staphylococcal Nuclease. 1977.

Smith, Margaret Meryl. Biochemistry of the Endosperm of *Lolium multiflorum Lam.* in Tissue Culture. 1972.

Soukchareun, Sommay. Preparation of Oligonucleotide: Peptide Hybrid Molecules as Antisense Inhibitors of Gene Expression. 1995.

Southwell, Bridget Rae. Synthesis of Plasma Proteins in Brain during Development. 1992.

Springell, Peter Henry. Amino Acid Metabolism with Special Reference to Peptide Bond Transfer. 1953.

Stanley, Peter George. Biochemical Studies on the Thyroid Gland, and; Observations on the Hydrolysis of Ovomucin. 1953.

Steel, Rohan. The Regulation of Stress Induced Apoptosis by Hsp72. 2006.

Stevens, Arthur. Studies on the Quaternary Structure of α-crystallin. 1989.

Stewart, Cameron Robert. Functional Roles of Serum Amyloid P Component in Amyloid Diseases. 2006.

Stewart, Peter Raymond. Physical and Biochemical Studies on the Proteins of Wheat. 1962.

Stone, Faye Helen. Studies on Resistance of Murine Leukemia Cells to cis-platinum. 1989.

Streader, L. Gail. Chemical and Structural Studies on the Lactate Dehydrogenase of *Streptococcus cremoris* us3. 1975.

Sullivan, John Joseph. Aspects of Proteolysis by Group N Streptococci. 1972.

Swan, J.M. Studies on the Chemistry of Proteins, Peptides and Amino Acids. (DSc) 1965.

Symons, Robert Henry. The Metabolism of Glucose and Fatty Acids by Tissues of the Sheep with Particular Reference to Ketone Body Formation. 1960.

Tan, Chor Teck. Targeted Analysis of Antigen Presentation Using Mass Spectrometry. 2012.

Teoh, Chai Lean. The Structure and Assembly of Apolipoprotein C-II Amyloid Fibrils. 2011.

Tesiram, Yasvir A. The Use of Adiabatic Pulses in NMR. 2003.

Thomson, John Allan. Alpha Crystallin: a Model for Multisubunit Protein Assemblies. 1985.

Thulborn, Keith Raymond. n-(9-Anthroyloxy) Fatty Acids as Fluorescent Probes for Biomembranes. 1979.

Todd, Pamela Ellen Emina. Proteolytic Enzymes of the Endocrine Glands. 1960.

Toulmin, Emma Louise. Investigating the Function of the Endogenous Prion Protein Using Genetic and Chemical Approaches. 2010.

Towns, Anthony Peter. Physicochemical Studies on Histones. 1970.

Trewhella, Maurice Arthur. The Measurement of Relative Turnover of Individual Molecular Species of Phospholipids. 1971.

Truscott, Roger John Willis. Changes in the Human Lens during Senile Nuclear Cataract Formation. 1976.

Tsykin, Anna. Gene Expression in Regenerating and Acute-phase Rat Liver and cDNA Sequence and Expression of Sheep Thyroxine-binding Globulin. 1993.

Tu, Eric. Regulatory T Cell Therapy for Late-stage Autoimmune Disease. 2011.

Turner, Bradley James. Toxic Mechanisms and Therapeutic Strategies in a Transgenic Mouse Model of Human Familial Amyotrophic Lateral Sclerosis. 2005.

Vadiveloo, Peter Krishnan. Characterisation of Apolipoproteins Involved in the Cellular Recognition of High Density Lipoproteins. 1990.

Van Laar, Ryan. Optimisation of cDNA Microarray Tumour Profiling and Molecular Analysis of Epithelial Ovarian Cancer. 2005.

Vella, Laura Jane. Investigating the Spread and Propagation of Infectious Prions. 2008.

Vince, James Edmund. Protein Trafficking and Organelle Biogenesis in *Leishmania* Parasites. 2005.

Walsh, Peter. Investigation of BEC: a Novel Ribosome Chaperone Complex. 2003.

Walters, Michael Thomas. The Role of Peroxidase in the Development of Stem Deformation in *Pinus radiata*. 1990.

Wan, David Chi-Cheong. Regulation of Opioid and Catecholamine Synthesis and Release in the Adrenal Medulla. 1989.

Wang, Yi. Developmental and Biochemical Studies of SOX13. 2005.

Ward, Alister Curtis. Studies on the Genes Responsible for Neurovirulence of Influenza A virus. 1994.

Ward, Jane Elizabeth. Pharmacology of Collateral Vessels. 1992.

Watson, D. Collected Papers – Contributions to Analytical and Medical Biochemistry. (DSc) 1963.

Weeks, Kate Louise. Targeting the IGF1-PI3K(p110α) Signalling Pathway to Improve Function of the Failing Heart. 2012.

Weidemann, Maurice John. The Metabolism of Fatty Acids by Animal Tissues. 1964.

Weiden, Sara. Hexosamine Metabolism: Under Normal and Pathological Conditions with Special Interest in Its Possible Bearing on Mucoprotein Deposition in Disease. 1956.

West, Alison Clare. Immunomodulatory Properties of the Histone Deacetylase Inhibitor Vorinostat. 2012.

Wiegmans, Adrian Peter. Investigating the Molecular Mechanisms of Apoptosis Targeted Therapy. 2010.

Whitley, Jane Carol. Genome Organisation in Mycoplasmas. 1991.

Wijayaratnam, Anne Primrose Wijayanthi. Construction and Analysis of Yeast Phospholipase C and Inositol Polyphosphate 5 Phosphatase Mutants. 2002.

Wilkinson, Tracey Nicole. Evolutionary Analysis of the Relaxin Peptide Family and Their Receptors. 2006.

Williams, Louise H. Identification of Novel Tumour Supressor Genes on Chromosome 22 Involved in Breast, Ovarian or Colon Cancer. 2007.

Williamson, Nicholas A. Characterisation and Methylation of Ribosomal Proteins. 2001.

Wong, Christina Sheau Fen. The Role of Siah2 in Breast Cancer Progression, Angiogenesis and Chemotherapeutic Responses. 2012.

Wong, Su Ee. Investigating the Significance of Relaxin in Experimental Diabetic Renal Disease. 2011.

Wong, Yuan Qi. Defining the In Vivo Kinetics of Mutant Huntingtin Aggregation in a Drosophila Model. 2013.

Wright, Simon Ward. Properties of Chloroplasts Isolated from Siphonous Green Algae. 1978.

Yang, Shuo. An Equilibrium Model for the Formation of apoC-II Amyloid Fibrils. 2012.

Yates, John Russell. The Oxidation of Thiol Groups and Their Relation to the Properties of Flour Proteins. 1960.

Ye, Siying. Molecular Characterization of Insulin-regulated Aminopeptidase (IRAP). 2006.

Youil, Rima. Molecular Biology of *Mycoplasma mycoides subsp. Mycoides*. 1991.

Young, Derek Geoffrey. The Side-chain Cleavage of Cholesterol and Cholesterol Sulphate by Enzymes from Adrenocortical Mitochondria. 1969.

Yu, Weiping. Isolation and Characterization of a Novel Phospholipid-activated Protein Kinase from Rat Liver. 1995.

Zawadzki, Jody Louise. The Biosynthesis of Novel Glycoconjugates in *Leishmania* Parasites. 2003.

Zhang, Jian-Guo. Structure-function Relationships of Interleukin-6. 1993.

Zhang, Qisen. Zoospore Differentiation and Potential Second Messenger Systems in the Fungus *Phytophthora palmivora*. 1990.

Zhang, Wei-zheng. Ageing of the Lens. 1993.

Zhang, Y. Mammalian Nuclear Proteins with DNA Binding Specificities Resembling MYB Proto-oncoproteins. 1995.

Zhao, Ling. A Study of the Physiological Effects of Relaxin: Generation and Analysis of Relaxin Deficient Mice. 2000.

Zhao, Tian. Stathmin, a Novel JNK Substrate. 2010.

Zhou, Xin-Fu. Regulation of Adrenal Medullary Secretion by Substance P. 1990.

Zwar, Tricia Deanne. Immunological Tolerance to Gastric Autoantigens. 2005.

Grimwade Prize Winners

The university's *Annual Report* for 1904/1905 announced the establishment of the Grimwade Prize, which derives from the gift by the father of Russell Grimwade, the Honourable Frederick Sheppard Grimwade (1840–1910). He gave £1000 for the foundation of an annual prize for the promotion of the study of industrial chemistry. The prize is open to any student who has devoted six months in the laboratories of the university to the investigation of some branch of industrial chemistry. It is not awarded every year. It has only once been won by a woman.

1907   Edwin Stuart Richards
Not awarded 1908–09
1910   Henry Caselli Richards
1911   Margaret Emilie Scott
1912   Kenneth Aubrey Mickle
1913   Royston Drew
Not awarded 1914–15
1916   Edward Ivan Rosenblum
Not awarded 1917–18
1919   Gustav Adolph Ampt
Not awarded 1920–23
1924   Neil Bannatyne Lewis
1925   Edwin John Roberts Drake
1926   Howard W. Strong
Not awarded 1927–32
1933   Ian William Wark

Not awarded 1934–36

1937   William Davies *and* William Gerard Jowett

Not awarded 1938–40

1941   John Stuart Anderson

1942   Erich Heymann *and* Abraham Yoffe

1943   Robert Alwyn Bottomley *and* Keith Leonard Sutherland

1944   Alan Wilson Wylie

Not awarded 1945

1946   F. Kenneth McTaggart

1947   Jack Norman Gregory

1948   Rupert Horace Myers

1949   Robert Lyle Croft

Not awarded 1950–54

1955   Donald Eric Weiss

1956   K.E. Murray

Not awarded 1957–61

1962   Ian Gordon McWilliam *and* Robert Alfred Dewar

Not awarded 1963–67

1968   Robert Alfred Dewar, V.E. Maier *and* Margaret A. Riddolls

Not awarded 1969–70

1971   Ian McKay Ritchie

Not awarded 1972–73

1974   Thomas William Healy

1975   Ian Edward Grey

1976   Robert Frederick Jung

Not awarded 1977–80

1981   David Roger Dixon

Not awarded 1982–83

1984   D. Neil Furlong

Not awarded 1985–87

1988   Franz Grieser

Not awarded 1989–91

1992   Martin G. Banwell

1993   Peter J. Scales

1994   John William Perich

Not awarded 1995–97

1998   Stephen Brown

1999   Paul Mulvaney

2000   Robert Capon

2001   David Dunstan

2002	Carl Schiesser
2003	Patrick Hartley
2004	Ewen Silvester
2005	Muthupandian Ashokkumar
2006	Anthony Dirk Stickland
2007	Roger Martin *and* Jonathan White
2008	Benjamin Boyd
2009	Bernard Luke Flynn
2010	Spencer John Williams
2011	Robert Lamb
2012	Spas Kolev
2013	Chris Burns

Sources Consulted

Books

Australian Academy of Science. *Science at the Shine Dome 2012: 100 Years of Antarctic Science, Program.* http://www.sciencearchive.org.au/events/sats/sats2012/documents/SATS2012-program.pdf

Bio21. *Annual Report.* 2004/05–2011/12.

Brady, Catherine. *Elizabeth Blackburn and the Story of Telomeres: Deciphering the Ends of DNA.* Cambridge, Mass.: MIT Press, 2007.

Dow, Hume (editor). *More Memories of Melbourne University: Undergraduate Life in the Years Since 1919.* Melbourne: Hutchinson, 1985.

Flesch, Juliet and Peter McPhee. *160 Years: 160 Stories: Brief Biographies of 160 Remarkable People Associated with the University of Melbourne.* Melbourne: Melbourne University Press, 2013.

Flesch, Juliet. *Life's Logic: 150 Years of Physiology at the University of Melbourne.* Melbourne: Australian Scholarly Press, 2012.

Flesch, Juliet. *Minding the Shop: People and Events that Shaped the Department of Property & Buildings 1853–2003 at the University of Melbourne.* Melbourne: Department of Property & Buildings, University of Melbourne, 2005.

Goad, Philip and George Tibbits. *Architecture on Campus: a Guide to the University of Melbourne and Its Colleges.* Melbourne: Melbourne University Press, 2003.

Humphreys, L.R. *Trikojus: a Scientist for Interesting Times.* Melbourne: Miegunyah Press, 2004.

McRae, Valda M. *Chemistry @ Melbourne 1960–2000.* Melbourne: School of Chemistry, University of Melbourne, 2007.

McRae, Valda M. *From Chalk and Talk to PowerPoint*. Melbourne: School of Chemistry, University of Melbourne, 2013.

Maxwell, Ivan. *Clinical Biochemistry*. Melbourne: Ramsay, 1925.

Maxwell, Ivan with the collaboration of members of the Departments of Biochemistry and Physiology of the University of Melbourne. *Clinical Biochemistry*. Melbourne: Melbourne University Press, 1956.

Penington, David. *Making Waves: Medicine, Public Health, Universities and Beyond*. Melbourne: Miegunyah Press, 2010.

Poynter, J.R. and Carolyn Rasmussen. *A Place Apart: the University of Melbourne, Decades of Challenge*. Melbourne: Melbourne University Press, 1996.

Poynter, J.R. *Russell Grimwade*. Melbourne: Miegunyah Press, 1967.

Radford, Joan. *The Chemistry Department of the University of Melbourne: Its Contribution to Australian Science, 1854–1959*. Melbourne: Hawthorn Press, 1978.

Russell, K.F. *The Melbourne Medical School, 1862–1962*. Melbourne: Melbourne University Press, 1977.

Selleck, R.J.W. *The Shop: the University of Melbourne, 1850–1939*. Melbourne: Melbourne University Press, 2003.

University of Melbourne. *Annual Report*. Melbourne: Melbourne University Press. 1889–. [1889–1976 published in the following year's *Calendar*.]

University of Melbourne. *Annual Report 1939–1946*. Melbourne: Melbourne University Press, 1948.

University of Melbourne. *Calendar*. Melbourne: Melbourne University Press, 1859–.

University of Melbourne. *List of Principal Benefactions Made to the University of Melbourne from Its Foundation in 1853 to the End of the Year 1957*. Melbourne: Melbourne University Press, 1959.

University of Western Australia School of Biomedical, Biomolecular and Medical Sciences: celebrating 95-50-5 years. Compiled by Evan Morgan, Howard Mitchell, Trevor Redgrave, Donald Robertson and Tony Bakker. Nedlands, 2007.

Who's Who of Australian Women. Compiled by Andrea Lofthouse based on research by Vivienne Smith. Sydney: Methuen Australia, 1982.

Journal Articles

'Another Win for Women'. *The Age*. 9 November 1971: 12.

Bebbington, Warren. 'Financial Support for the Collections in 2006 and 2007'. *University of Melbourne Collections*. v. 1 (November 2007): 48–9.

Bebbington, Warren. 'Introduction'. *University of Melbourne Collections*. v. 2 (July 2008): 2.

Benjamin, Jason. 'Amateur Perfection: the Photograph Albums of Sir Russell Grimwade'. *University of Melbourne Library Journal*. 2005: 15–22.

Breinl, Anton and W.J. Young. 'Tropical Australia and Its Settlement'. *Annals of Tropical Medicine and Parasitology*. v. 13 (1920): 351–412.

'Cahn-Osborne Wedding'. *Argus*. 13 July 1929: 12.

'Clarke, Adrienne Elizabeth (1938–)'. http://www.eoas.info/biogs/P002212b.htm

Cold Spring Harbor Laboratory Library. *Mary-Jane Gething Oral History Interview, Cold Spring Harbor Laboratory Library 16 January 2003*. http://library.cshl.edu/oralhistory/category/cshl/

Cole, Barry L. 'Robert Augusteyn: Director of the National Vision Research Institute of Australia 1991 to 2001'. *Clinical and Experimental Optometry*. v. 85 no.2 (March 2002): 107–10.

Crane, Denis and Marie Bogoyevitch. 'Great Expectations'. *Australian Biochemist*. v. 38 no. 1 (April 2007): 28–30.

Ebert, David. 'Great Expectations'. *Australian Biochemist*. v. 35 no. 3 (December 2004): 26–9.

Fennessy, Joe. 'New Partnerships Support Indigenous Health'. *Voice*. v. 10 no. 1 (13 January – 9 February 2014).

Flesch, Juliet. '"A Biochemist of the Best Type": the Contribution of Arthur Cecil Hamel Rothera to Biochemistry in Australia'. *Historical Records of Australian Science*. v. 23 (2012): 120–31.

Flesch, Juliet. 'The Ones that Got Away: Four Women from the Department of Physiology and What They Did Next'. *University of Melbourne Collections*. Issue 11 (December 2012): 44–50.

Gleeson, Paul. 'Farewell to the Russell Grimwade School of Biochemistry Building, University of Melbourne'. *Australian Biochemist*. v. 39 no. 2 (August 2008): 28–39.

Gottschalk, A. and K. Law. 'Professor William John Young, DSc (Lond.), MSc (Manchester) 1878–1942'. *Science Review*. August 1942: 2–6.

Hannink, Nerissa. 'Malaria Goes Bananas before Reproduction'. *Voice*. v. 8 no. 3 (12 March – 8 April 2012).

Hird, Frank and Max Marginson. 'The Other Side of a Scientist's Story'. *Sydney Morning Herald*. 24 August 1999.

Hird, F.J.R., A.L. Lazer and W.K. Tickner. 'W.A. Rawlinson'. *University Gazette*. August 1972: 14–15.

Hoogenraad, Nick. 'In Memoriam: Bruce Arthur Stone (1928–2008)'. *Australian Biochemist*. v. 39 no. 3 (December 2008): 28–9.

Hopkins, F. Gowland. 'Arthur Cecil Hamel Rothera'. *Biochemical Journal*. v. 10 no. 1 (March 1916): 11–13.

Hume, E.M. 'Obituary: Charles James Martin, Kt, C.M.G., F.R.C.P., D.Sc., F.R.S.

(9 January 1866 – 15 February 1955)'. *British Journal of Nutrition*. v. 10 no. 1 (1956): 1–7.

Legge, J.W. and F. Gibson. 'Victor Martin Trikojus, 1902–1985'. *Historical Records of Australian Science*. v. 6 no. 4 (July 1987): 519–31.

Marginson, Ray. 'Days of Wine and Eucalypts and the Chemistry of Living: Maxwell Arthur Marginson'. *Age*. 22 October 2002: 11.

Mark VK3PI. 'Vale Robert (Bob) Goullet, VK3BU'. *WANSARC News*. v. 40 no. 4 (May 2009): 2.

'Medical Students Vote to Identify Teaching Excellence'. *UniNews*. v.12 no. 6 (21 April – 5 May 2003). http://archive.uninews.unimelb.edu.au/news/550/

Melbourne Newsroom, 12 September 2012. '"Mad Cow" Blood Test Now on the Horizon'. http://newsroom.melbourne.edu/news/n-901

'Melbourne Red Scientist at Commission'. *Sydney Morning Herald*. 30 October 1954: 5.

Moore, Colin E. 'Herbert Frank Shorney MD, FRCS (1878 – 2 May 1933)'. *Australian Journal of Ophthalmology*. v. 12 (1984): 289–91.

Moore, Winifred. 'Girls Take University Seriously'. *Courier-Mail*. 31 August 1949.

Moses, Ken. 'Why Keep It Quiet? They Can't Miss This'. *Argus*. 9 February 1955: 2.

Mulcahy, Elaine. 'Leading Dental Research Wins Uni Scientist Victoria Prize'. *UniNews*. v. 14 no. 15 (22 August – 5 September 2005).

Osborne, W.A. 'Obituary: William John Young'. *Medical Journal of Australia*. June 1942: 707–8.

Parslow, Graham R. 'Multimedia in Biochemistry and Molecular Biology Education, Commentary: Massive Open Online Courses'. *Biochemistry and Molecular Biology Education*. v. 41 no. 4 (August 2013): 278–9.

'Researchers to Benefit from a Share of More than $8.1 Million Awarded in Latest Round of NHMRC funding'. 29 October 2013. http://www.bio21 .unimelb.edu.au/news/researchers-to-benefit-from-a-share-of-more-than-8-1-million-awa

Ryle, Gerald and Gary Hughes. 'How Australia Raided the Great Minds of Hitler's War Machine'. *Sydney Morning Herald*. 16 August 1999; also published in *The Age*.

Scott, Rebecca. 'Elizabeth Blackburn School of Sciences Launched'. *Voice*. v. 10 no. 4 (14 April – 11 May 2014): 6.

Skerritt, Henry. 'Making the Past Present in the Sir Russell and Lady Mab Grimwade Collection'. *University of Melbourne Collections*. Issue 10 (June 2012): 15–23.

Stone, Bruce. 'Reflections'. *Australian Biochemist*. v. 31 no. 4 (December 2000): 19–20.

Sweet, M. 'The Profile: Kerin O'Dea'. *Australian Doctor*. 2000: 27–9.

'Teaching Good Value'. *Voice*. v. 5 no. 1 (13 April – 10 May 2009). http://archive.uninews.unimelb.edu.au/news/5772/

'$10,000 Grant for Cinnamon Fungus Research'. *University Gazette*. March 1981: 14.

Tilley, Leann and Bruce Stone. 'A Scientist Ahead of Her Times: Audrey Josephine Cahn: Nutritionist, Artist, 17.10.1905-1.4.2008'. *The Age*. 12 May 2008.

Trikojus, V.M. 'Biochemistry at the University of Melbourne'. *Nature*. v. 191 no. 4795 (23 September 1961): 1238–40.

Trikojus, V.M. 'Leslie Algernon Ivan Maxwell'. *Medical Journal of Australia* (1965): 256–7, 509–10.

Trikojus, V.M. 'Robert Kerford Morton'. *Australian Academy of Science Yearbook*. 1954.

University of Melbourne. *Staff News*. 17 August 1970: 6.

Vaux, David. 'Interviews with Australian Scientists: Professor Nick Hoogenraad, Biochemist'. http://sciencearchive.org.au/scientists/interviews/h/nhoogenraad.html

Ward, H.A. 'Fantl, Paul (1900–1972)'. *Australian Dictionary of Biography*. Melbourne: Melbourne University Press, 1996.

Weaver, Chris. 'Record Gift to Fund Art Conservation Chair and Centre'. *Voice*. v. 10 no. 4 (14 April – 11 May 2014): 16.

Whitton, W.I. 'Robert Alfred Dewar, 1908–1981'. *Chemistry in Australia*. v. 48 no. 4 (April 1981): 164.

Archival Sources

University of Melbourne Archives. F.R.J. Hird. Notes on Victor Trikojus.

University of Melbourne Archives. *Registrar's Correspondence*.

University of Melbourne Archives. Trikojus papers.

University of Melbourne Council. *Minutes*.

Online sources

Advance. 'Dr Russell Howard'. http://advance.org/russell-howard/

Caples, Amanda. 'Melbourne Researchers Close in on "Mad Cow" Blood Test'. Invest Victoria. 5 October 2012. http://blog.invest.vic.gov.au/2012/10/05/melbourne-researchers-close-in-on-mad-cow-blood-test/

Carey, Jane. '"What's a Nice Girl Like You Doing with a Nobel Prize?" Elizabeth Blackburn, "Australia's" First Woman Nobel Laureate and Women's Scientific Leadership'.

http://www.womenaustralia.info/leaders/sti/pdfs/19_Carey.pdf

Florigene Pty Ltd. http://www.florigene.com/?country=Australia

La Trobe University. 'Colorectal Cancer Secretome & Exosome Biology. Simpson Lab'. http://www.latrobe.edu.au/biochemistry/specialisations/simpson

Millis, Nancy. Professor Nancy Millis interviewed by Ms Sally Morrison in 2001. *Interviews with Australian Scientists*. Australian Academy of Science. http://www.sciencearchive.org.au/scientists/interviews/m/nm.html

Outback Stores. http://outbackstores.com.au/

Tilley, Leann. 'Audrey Cahn: a Nonagenarian Scientist Remembers the Early Days'. WiSENet. Issue 49.

Notes

Introduction

1   J.R. Poynter. *Russell Grimwade*. Melbourne: Miegunyah Press, 1967.

2   Biographical notes on the Grimwade family may be found in the various volumes of the *Australian Dictionary of Biography*.

3   http://www.trinity.unimelb.edu.au/learning/centre-for-advanced-studies/endowed-visiting-lectureships/grimwade-lectureships.html

4   University of Melbourne. *List of Principal Benefactions Made to the University of Melbourne from Its Foundation in 1853 to the End of the Year 1957*. Melbourne: Melbourne University Press, 1959.

5   P.R.H. St John worked for fifty-one years in the Melbourne Botanic Gardens and studied under Ferdinand von Mueller. His death at seventy-three was recorded in the *Argus* edition of 14 August 1944.

6   http://www.botany.unimelb.edu.au/buffalo/about.htm

7   UTR7.235 and UTR7. 236.

8   A list can be found at http://msl.unimelb.edu.au/learning-teaching/miegunyah/fellows.

9   Chris Weaver. 'Record Gift to Fund Art Conservation Chair and Centre'. *Voice*. v. 10 no. 4 (14 April – 11 May 2014):16.

10 *The Grimwade Collection: a Selection of Works from the Bequest of Sir Russell and Lady Grimwade: University Gallery, the University of Melbourne, 22 July – 4 September 1987*. Melbourne: University Gallery, University of Melbourne, 1987; *Works of Art from the Russell and Mab Grimwade Bequest: the University of Melbourne Art Collection*. Melbourne: Museum of Art, University of Melbourne, 1989; Lisa Sullivan. *A Collection and a Cottage*. Melbourne: Ian Potter Museum of Art, The University of Melbourne, 2000.

Chapter 1— In the Beginning

1 Juliet Flesch. *Life's Logic: 150 Years of Physiology at the University of Melbourne*. Melbourne: Australian Scholarly Publishing, 2012.

2 Joan Radford. *The Chemistry Department of the University of Melbourne: Its Contribution to Australian Science, 1854–1959*. Melbourne: Hawthorn Press, 1978. pp. 14–15.

3 K.F. Russell. 'Macadam, John (1827–1865)'. *Australian Dictionary of Biography*. Melbourne: Melbourne University Press, 1974.

4 'The Late Dr Macadam'. *Australian Medical Journal*. October 1865: 327–32.

5 K.F. Russell. *The Melbourne Medical School 1862–1962*. Melbourne: Melbourne University Press, 1977. p. 38.

6 University of Melbourne. *Calendar*. 1866–67. p. 143.

7 R.J.W. Selleck. *The Shop: the University of Melbourne, 1850–1939*. Melbourne: Melbourne University Press, 2003. pp. 196–9.

8 Radford. *The Chemistry Department of the University of Melbourne*.

9 W.A. Osborne. *Elementary Practical Biochemistry*. Melbourne: Ramsay, 1920.

10 University of Melbourne Council. *Minutes*. 26 June 1905.

11 Ibid.

12 Juliet Flesch. '"A Biochemist of the Best Type": the Contribution of Arthur Cecil Hamel Rothera to Biochemistry in Australia'. *Historical Records of Australian Science*. v. 23 (2012): 120–31.

13 F. Gowland Hopkins. 'Arthur Cecil Hamel Rothera'. *Biochemical Journal*. v. 10 no. 1 (March 1916): 11–13.

14 Selleck. *The Shop*. p. 488.

15 'The Use of Ammonium Salts in Biochemistry'. 18 May 1909; 'Physical Properties of Gels'. 5 July 1910.

16 A.C.H. Rothera. 'Note on the Sodium Nitro-Prusside Reaction for Acetone'. *Journal of Physiology*. v. 37 (1908): 491–4.

17 See, for example, B.B. Tripathy, editor-in-chief. *RSSDI Textbook of Diabetes Mellitus*. 2nd rev. ed. New Delhi: Jaypee, 2012; Oluwafemi S. Obayori, Matthew O. Ilori, Sunday A. Adebusoye, Ganuyu O. Oyetibo and Olukayonde O. Amund. 'Pyrene-Degradation Potentials of Pseudomonas Species Isolated from Polluted Tropical Soils'. *World Journal of Microbiology and Biotechnology*. v. 24 no. 11 (November 2008): 2639–46.

18 A list of Rothera's papers appears in W.A. Osborne. 'The Late Captain A.C.H. Rothera'. *Speculum*. October 1915: 260–3; and Juliet Flesch. 'A Biochemist of the Best Type'.

19 University of Melbourne Archives (UMA). *Registrar's Correspondence*. 1913/314.

20 I am indebted to Dr Peter Hobbins for clarifying in personal correspondence that Rothera did not die of meningitis which was present in the hospital.

21 'Funeral of Captain Rothera'. *Argus*. 5 October 1915: 8.

22 Lilias Maxwell to Trikojus. 17 April 1958. UMA. Trikojus papers.

23 W.A. Osborne and L.C. Jackson. 'Counter Diffusion in Aqueous Solution'. *Biochemical Journal*. v. 8 no. 3 (June 1914): 246–9.

24 Registrar's correspondence. 1915.

25 V.M. Trikojus. *Medical Journal of Australia*. 1965: 256–7, 509–10.

26 Ivan Maxwell. *Clinical Biochemistry*. Melbourne: Ramsay, 1925; Ivan Maxwell with the collaboration of members of the Departments of Biochemistry and Physiology of the University of Melbourne. *Clinical Biochemistry*. Melbourne: Melbourne University Press, 1956.

27 Richard Travers. 'Wilkinson, John Francis (1864–1935)'. *Australian Dictionary of Biography*. Melbourne: Melbourne University Press, 1990. v. 12.

28 A. Gottschalk and K. Law. 'Professor William John Young, DSc (Lond.), MSc (Manchester) 1878–1942'. *Science Review*. August 1942: 2–6.

29 W.A. Osborne. 'Obituary: William John Young'. *Medical Journal of Australia*. June 1942: 707–8.

30 Anton Breinl and W.J. Young. 'Tropical Australia and Its Settlement'. *Annals of Tropical Medicine and Parasitology*. v. 13 (1920): 351–412.

31 R.A. Douglas. 'Breinl, Anton (1890–1944)'. *Australian Dictionary of Biography*. Melbourne: Melbourne University Press, 1979. See also Osborne, 'Obituary: William John Young'.

32 Gottschalk and Law. 'Professor William John Young, DSc (Lond.), MSc (Manchester) 1878–1942'.

33 'Let's Talk of Interesting People'. *Australian Women's Weekly*. 10 April 1937: 2; 'African Mothers "Go Bush"'. *Barrier Miner*. 3 January 1951: 7.

Chapter 2—The War Years

1 Barry O. Jones. 'Osborne, William Alexander (1873–1967)'. *Australian Dictionary of Biography*. Melbourne: Melbourne University Press, 1988. v. 11.

2 F. Gowland Hopkins. 'Arthur Cecil Hamel Rothera'. *Biochemical Journal*. v. 10 no. 1 (March 1916): 11–13.

3 University of Melbourne. *Annual Report 1939–1946*. Melbourne: Melbourne University Press, 1948.

4 L.R. Humphreys. *Trikojus: a Scientist for Interesting Times*. Melbourne: Miegunyah Press, 2004.

5 University of Melbourne. *Annual Report 1939–1946*. p. 137.

6 Ibid.

7 J.W. Legge and F. Gibson. 'Victor Martin Trikojus, 1902–1985'. *Historical Records of Australian Science*. v. 6 no. 4 (July 1987): 519–31.

8 Juliet Flesch and Peter McPhee. 'Moshi (Mowsey) Inagaki (1880–1947)'. *160 Years: 160 Stories: Brief Biographies of 160 Remarkable People Associated with the University of Melbourne*. Melbourne: Melbourne University Press, 2013. p. 87; Gerald Ryle and Gary Hughes. 'How Australia Raided the Great Minds of Hitler's War Machine'. *Sydney Morning Herald*. 16 August 1999. Also published in *The Age*.

9 Frank Hird and Max Marginson. 'The Other Side of a Scientist's Story'. *Sydney Morning Herald*. 24 August 1999.

10 UMA. F.R.J. Hird. Notes on Victor Trikojus.

11 UMA. Trikojus papers.

12 Ibid.

13 Ibid.

14 Juliet Flesch. *Minding the Shop: People and Events that Shaped the Department of Property & Buildings 1853–2003 at the University of Melbourne*. Melbourne: Department of Property & Buildings, University of Melbourne, 2005. pp. 82–95.

15 UMA. Trikojus papers.

16 University of Melbourne Council. *Minutes*. 5 April 1943; UMA. Trikojus papers.

17 Legge and Gibson. 'Victor Martin Trikojus, 1902–1985'.

18 V.M. Trikojus and Muriel G. Crabtree. 'The Thyrotrophic Hormone, Thyroxine and Ascorbic Acid in Relationship to the Liver Glycogen of Guinea-Pigs'. *Biochemical Journal*. 1946; V.M. Trikojus and Mary McQuillan. 'The Thyrotrophic Hormone and Thiourea in Reference to the Problem of 'Antihormones'. *British Journal of Experimental Pathology*. v. 27 no. 247 (1946); V.M. Trikojus and Lesbia E.A. Wright. 'The Reaction of Iodine with Preparations of the Thyretrophic Hormone'. *Medical Journal of Australia*. 1946; V.M. Trikojus. 'Chemistry of Sodium Ethylmercurithiosalicyclate'. *Nature*. 1946.

19 Valda M. McRae. *From Chalk and Talk to PowerPoint*. Melbourne: School of Chemistry, University of Melbourne, 2013.

20 D.W. Cameron, W.H. Sawyer and V.M. Trikojus. 'Colouring Matters of the Aphidoidea XLII. Purification and Properties of the Cyclising Enzyme (Protoaphin Dehydratase (Cyclising)) Concerned with Pigment Transformations in the Woolly Aphid *Eriosoma lanigerum* Hausmann (Hemiptera: Insecta)'. *Australian Journal of Biological Sciences*. v. 30 no. 3 (1977): 173–82.

21 http://www.csiropedia.csiro.au/display/CSIROpedia/Home

22 Personal correspondence. April 2013.

23 UMA. Trikojus papers.

24 Humphreys. *Trikojus*. p. 98.

25 UMA. Trikojus papers.

26 Ibid.

27 UMA. Trikojus papers. Trikojus to Kathleen Law.

28 UMA. Trikojus papers. Turner to Trikojus. 14 August 1951; Trikojus to Turner. 11 October 1951.

Chapter 3—Trailblazing Women

1 Robert Robison, Kathleen Alice O'Dell Law and Adèle Helen Rosenheim. 'Deposition of Strontium Salts in Hypertrophic Cartilage in Vitro'. *Biochemical Journal*. v. 30 (January 1936): 66–8; Kathleen Alice O'Dell Law and Robert Robison. 'The Influence of Changes Induced by Cholesterol upon the Calcification in Vitro of Rabbit Aorta'. *Biochemical Journal*. v. 30 (January 1936): 69–75.

2   UMA. Trikojus papers.

3   Kathleen Law. 'Phenol Oxidases in Some Wood-rotting Fungi'. *Annals of Botany*.
    nS v. 14 no. 53 (January 1956): 69–78; Kathleen Law. 'Laccase and Tyrosinase in
    Some Wood-rotting Fungi'. *Annals of Botany*. nS v. 19 no. 76 (1955): 561–70.

4   Flesch. *Life's Logic*. pp. 95–7.

5   Dr Jean Jackson interviewed by Juliet Flesch. 28 October 2010.

6   UMA. Trikojus papers.

7   Ibid.

8   Ibid.

9   Professor Nancy Millis interviewed by Ms Sally Morrison in 2001. *Interviews with
    Australian Scientists*. Australian Academy of Science. http://www.sciencearchive.
    org.au/scientists/interviews/m/nm.html

10 Jean Millis and You Poh Seng. 'The Effect of Age and Parity of the Mother on
    Birth Weight of the Offspring'. *Annals of Human Genetics*. v. 19 (1954): 58–73.

11 UMA. Trikojus papers.

12 Juliet Flesch. 'The Ones that Got Away: Four Women from the Department
    of Physiology and What They Did Next'. *University of Melbourne Collections*.
    Issue 11 (December 2012): 44–50.

13 'Cahn-Osborne Wedding'. *Argus*. 13 July 1929: 12.

14 *Argus*. 16 August 1930: 13.

15 Audrey Cahn. 'The Dietary Department of the Out-Patient Section of a Public
    Hospital'. Melbourne, 1937.

16 Leann Tilley. 'Audrey Cahn: a Nonagenarian Scientist Remembers the Early
    Days'. Wisenet. Issue 49 (November 1998). http://www.wisenet-australia.org/

17 Kilvington to registrar. 14 May 1947. Audrey Cahn personnel file.

18 Frank L. Apperley and Muriel G. Crabtree. 'Relation of Gastric Function to the
    Chemical Composition of the Blood'. *Journal of Physiology*. 1931.

19 Muriel G. Crabtree and Montague Maizels. 'The Sodium Content of Human
    Erythrocytes'. *Biochemical Journal*. v. 31 no. 12 (December 1937): 2153–4.

20 Winifred Moore. 'Girls Take University Seriously'. *Courier-Mail*. 31 August 1949.

21 Dora Winikoff, 'Calcium, Magnesium and Phosphorus in the Milk of Australian
    Women'. *Medical Journal of Australia*. 1944.

22 Dora Winikoff and V.M. Trikojus. 'N^1-Diethylsulphanilamide: a Reagent for the
    Colorimetric Estimation of Thyroxine'. *Biochemical Journal*. v. 2 no. 3 (1948):
    475–80.

23 Dora Winikoff and Malvina Malinek. 'The Predictive Value of Thyroid "Test
    Profile" in Habitual Abortion'. *British Journal of Obstetrics and Gynaecology*. v. 82
    (September 1975): 760–6.

24 Humphreys. *Trikojus*. p. 98.

25 University of Melbourne Council. *Minutes*. 4 May 1959.

26 UMA. Trikojus to the registrar. 15 June 1960.

27 McQuillan personnel record. 23 January 1974.

28 Mary T. McQuillan, *Somatostatin*. Edinburgh: Churchill Livingstone, 1977–79.

Chapter 4—After the War

1 H.A. Ward. 'Fantl, Paul (1900-1972)'. *Australian Dictionary of Biography*. Melbourne: Melbourne University Press, 1996.

2 UMA. Trikojus papers.

3 F.J.R. Hird, A.L. Lazer and W.K. Tickner. 'W.A. Rawlinson'. *University Gazette*. August 1972: 14–15.

4 Ibid.

5 Ken Moses. 'Why Keep It Quiet? They Can't Miss This'. *Argus*. 9 February 1955: 2.

6 Evan Morgan, Howard Mitchell, Trevor Redgrave, Donald Robertson and Tony Bakker (compilers). *University of Western Australia School of Biomedical, Biomolecular and Medical Sciences: Celebrating 95-50-5 Years*. Nedlands, 2007.

7 UMA. Trikojus papers.

8 Ibid.

9 Ibid.

10 University of Melbourne. *Annual Report 1939–1946*. p. 153.

11 Ibid.

12 Ibid.

13 Ibid. Heinz August Kamphausen (1907–1964) was in charge of the first glass-blowing workshop in the Department of Chemistry, University of Melbourne in 1949; the workshop continued until 1964.

14 P.H. Springell. 'For the Freedom to Comment by Scientists'. *Arena*. Issue 44/45 (1976): 28–33.

15 UMA. Trikojus papers.

16 University of Melbourne. *Annual Report 1939–1946*. p. 110.

17 F.J.R. Hird and V.M. Trikojus. 'Paper Partition Chromatography with Thyroxine and Analogues'. *American Journal of Botany*. v. 35 no. 7 (July 1948): 185–7.

18 *University Gazette*. November 1964: 7.

19 UMA 1986:0065. Hird papers.

20 University of Melbourne Academic Board. *Minute of Appreciation: Francis John Raymond Hird*. 18 December 1985.

21 Frank Hird and Max Marginson. 'The Other Side of a Scientist's Story'.

22 University of Melbourne Council. *Minutes*. 6 November 1961. For details of the publication, see F.J.R. Hird. 'The Reduction of Serum Albumin, Insulin and Some Simple Disulphides by Glutathione'. *Biochemical Journal*. v. 85 no. 2 (November 1962): 320–6.

23 Ibid.

24 R. Lemberg and J.W. Legge. *Hematin Compounds and Bile Pigments: Their Constitution, Metabolism and Function*. New York: Interscience, 1949.

25 Max Marginson. 'Jack Legge: an Appreciation'. *Newsletter*. Australian Society for Biochemistry and Molecular Biology Inc. v. 28 no.1 (March 1997): 16–17.

26 UMA. Trikojus papers.

27 Ibid.
28 'Melbourne Red Scientist at Commission'. *Sydney Morning Herald*. 30 October 1954: 5.
29 University of Melbourne Council. *Minutes*. 4 May 1959.
30 Marginson. 'Jack Legge'.
31 J.W. Legge, 'Presidential Address, 1950: the Origin of Life'. *Science Review*. Issue 13 (1950–51): 38–43; A.W. Turner and J.W. Legge. 'Bacterial Oxidation of Arsenite. II. The Activity of Washed Suspensions'. *Australian Journal of Biological Sciences*. v. 7, no. 49 (November 1954): 479–95; J.W. Legge and A.W. Turner. 'Bacterial Oxidation of Arsenite. III. Cell-Free Arsenite Dehydrogenase'. *Australian Journal of Biological Sciences*. v. 7, no. 49 (November 1954): 496–503; J.W. Legge. 'Bacterial Oxidation of Arsenite. IV. Some Properties of the Bacterial Cytochromes'. *Australian Journal of Biological Sciences*. v. 7, no. 49 (November 1954): 504–14; P. Caligiore, F. Macrae, D. John, L. Rayner and J.W. Legge. 'Peroxidase Levels in Food: Relevance to Colorectal Cancer Screening'. *American Journal of Clinical Nutrition*. no. 35 (June 1982): 1487–9.
32 V.M. Trikojus. 'Robert Kerford Morton'. *Australian Academy of Science Yearbook*. 1954. Also published as a pamphlet and accessible online at http://sciencearchive.org.au/fellows/documents/morton.pdf
33 Humphreys. *Trikojus*. p. 100.
34 Graham Parslow. 'Then and Now: Biochemical Education'. *Australian Biochemist*. v. 6 no. 2 (August 2005): 58.
35 R.K. Morton. 'Separation and Purification of Enzymes Associated with Insoluble Particles'. *Nature*. v. 166 (December 1950): 1093–5.
36 Trikojus. 'Robert Kerford Morton'.
37 J.E. Falk, R. Lemberg and R.K. Morton (editors). Symposium on Haematin Enzymes. *Haematin Enzymes: a Symposium of the International Union of Biochemistry, Organized by the Australian Academy of Science*. Oxford; New York: Pergamon Press, 1961.
38 R.K. Morton. 'New Concepts of the Biochemistry of the Cell Nucleus'. *Australian Journal of Science*. December 1961: 260–78.
39 *Research Report*. 1966. p. 220.
40 Harris Busch. 'The Final Common Pathway of Cancer: Presidential Address'. *Cancer Research*. v. 50 (15 August 1990): 4830–8.
41 Hume Dow (editor). *More Memories of Melbourne University: Undergraduate Life in the Years Since 1919*. Melbourne: Hutchinson, 1985. pp. 89–117.
42 Ray Marginson. 'Days of Wine and Eucalypts and the Chemistry of Living: Maxwell Arthur Marginson'. *Age*. 22 October 2002: 11.
43 F.J.R. Hird and M.A. Marginson. 'Formation of Ammonia from Glutamate by Mitochondria'. *Nature*. v. 210 no. 4925: 1224–5; F.J.R. Hird and M.A. Marginson. 'The Formation of Ammonia from Glutamine and Glutamate by Mitochondria from Rat Liver and Kidney'. *Archives of Biochemistry and Biophysics*. v. 127 (1968): 718–24.

44 M.A. Marginson. 'The Science Faculty Bureau'. *Science Review*. 1948: 35–7.

45 'Dietary Deficiencies in the Supermarket Society'. *University Gazette*. December 1983: 3–4.

46 J.S. Rogers. *The University of Melbourne Mildura Branch, 1947–1949: a Short History*. Edited and illustrated by Norman H. Olver. Melbourne: The University of Melbourne Mildura Branch 40th Anniversary Reunion Committee, 1991. p. 125.

47 University of Melbourne Council. *Minutes*. 1978.

48 Russell. *The Melbourne Medical School, 1862–1962*. p. 197.

49 UMA. Trikojus papers.

50 Ibid.

51 Ibid.

52 Flesch. *Life's Logic*. pp. 29–30.

53 UMA. Trikojus papers.

54 Ibid.

55 Flesch. *Minding the Shop*. pp. 82–92.

56 UMA. Trikojus papers.

Chapter 5—The Great Shift

1 Poynter. *Russell Grimwade*.

2 Registrar's correspondence V-C 163.

3 University of Melbourne Council. *Minutes*. 2 October 1950.

4 Chancellor's report to University of Melbourne Council. 1955.

5 Flesch. *Minding the Shop*. pp. 187–8.

6 Now the David Jones store: the original name appears above the canopy.

7 Marginson. 'Jack Legge: an appreciation': 16–17.

8 Philip Goad and George Tibbits. *Architecture on Campus: a Guide to the University of Melbourne and Its Colleges*. Melbourne: Melbourne University Press, 2003. p. 63.

9 University of Melbourne Council. *Minutes*. 7 November 1955.

10 University of Melbourne Council. *Minutes*. 17 February 1958.

11 *Report of the Committee on Australian Universities, September 1957*. Canberra: Government Printer, 1958.

12 University of Melbourne Council. *Minutes*. 6 November 1961.

13 UMA. V-C/163. Undated draft letter (1961) for V-C's signature apparently by Trikojus, evidently not sent, as there is a note by Paton to Nessie Rennie saying 'Pls get RALTON (Bates, Smart & McCutcheon on the phone)'.

14 UMA. Trikojus papers.

15 Ibid.

16 'Centenary Woollen Mills'. *Frankston Standard*. 9 April 1943: 3; 'Australian Knitwear Undersold'. *Adelaide Advertiser*. 22 June 1950: 3; '200 Dismissed by Vic. Woollen Mill'. *Brisbane Courier-Mail*. 3 November 1951: 1.

17   S.J. Leach and E.M.J. Parkhill. 'Amide Nitrogen in Proteins'. International Wool Textile Research Conference. *Proceedings*. Melbourne: CSIRO, 1956.

18   See, for example, Robert C. Augusteyn, Evelyn M. Parkhill and Arthur Stevens. 'The Effect of Isolation Buffers on Properties of α-Crystallin'. *Experimental Eye Research*. v. 54 (1992): 219–28); S.C. Cianciosi and Francis J.R. Hird. 'The Collagen Content of Selected Animals'. *Comparative Biochemical Physiology*. v. 85B no. 2 (1986): 295–8.

19   Peter F. Hall. *The Physician and the Biochemist*. Melbourne: P. Hall, 1971.

20   A.J. Farnworth. 'A Hydrogen Bonding Mechanism for the Permanent Setting of Wool Fibers'. *Textile Research Journal*. v. 27 no. 8 (August 1957): 632–40.

21   'SiroSet'. *CSIROpedia: Telling the Story of CSIRO's Achievements and Achievers*. http://www.csiropedia.csiro.au/display/CSIROpedia/SiroSet

22   Readers interested in Leach's recollections of his early career should consult 'Reflections, Syd Leach: Chemical Engineering to Protein Folding, Syd Leach Reflects on His Early Work in the UK then at CSIRO in Melbourne'. *Australian Biochemist*. v. 31 no. 2 (August 2000): 18–19; v. 31 no. 4 (December 2000): 21–3.

23   C.A. Appleby, N.A. Nicola, J.G.R. Hurrell and S.J. Leach. 'Characterization and Improved Separation of Soybean Leghemoglobins'. *Biochemistry*. v. 14 no. 20 (1975): 4444–50; N.A. Nicola, E. Minasian, C.A. Appleby and S.J. Leach. 'Circular Dichroism Studies of Myoglobin and Leghemoglobin'. *Biochemistry*. v. 14 no. 23 (1975): 5141–9.

24   'The "Prof" was Pioneer in Biochemistry'. *Newcastle Herald*. 14 February 2005: 53.

25   S.J. Leach. 'The Evolution of Proteins'. *Newtrino*. v. 7 no. 1 (June 1976): 44–9.

26   Maurie Trewhella and Bridget Underhill. 'Frederick Darien Collins'. https://www.asbmb.org.au/fredcollins.html

27   For example, F.D. Collins and R.A. Morton. 'Absorption Spectra, Molecular Weights and Visual Purple'. *Nature*. v. 164 no. 4169 (24 September 1949): 528–9; F.D. Collins and R.A. Morton. 'Retinal Receptors'. *Nature*. v. 167 no. 4252 (28 April 1951): 673–4.

28   Trewhella and Underhill. 'Frederick Darien Collins'; F.D. Collins. 'Phospholipids Containing Phosphate Triester Groups'. *Nature*. Issue 188 (22 October 1960): 297–300.

29   A.I. Feher, F.D. Collins and T.W. Healy. 'Mixed Monolayers of Simple Saturated and Unsaturated Fatty Acids'. *Australian Journal of Chemistry*. v. 30 no. 3 (1977): 511–17.

30   Academic Board. *Minute of Appreciation*. 18 December 1997.

31   *Age*. 14 March 1993.

32   University of Melbourne staff file.

33   Academic Board. *Minute of Appreciation*. 24 August 2000.

34   B.E. Davidson, M. Sajgò, H.F. Noller and J. Ieuan Harris. 'Amino Acid Sequence of Glyceraldehyde 3-Phosphate Dehydrogenase from Lobster Muscle'. *Nature*. v. 216 (23 December 1967): 1181–5.

35 'Bacteria Enlisted to Make Natural Food Preservatives'. *UniNews*. 12 July 1996: 3.

36 Bill Sawyer. 'An Unplanned Biochemical Life: Bill Sawyer Reflects on His Adventures in Biochemistry in Australia and Overseas'. *Australian Biochemist*. v. 31 no. 1 (April 2000): 23–4.

37 Donald J. Winzor and William H. Sawyer. *Quantitative Characterization of Ligand Binding*. New York: Wiley-Liss, 1995.

38 '1980 David Syme Research Prize'. *University Gazette*. v. 37 no. 1 (March 1981): 13.

39 Anne McDougall, W.H. Sawyer and V. Ciesielski. *A Computer-based Simulation Exercise in Biochemistry*. Melbourne: Computers in Education Research Group, University of Melbourne, 1974.

40 UMA. Trikojus papers. Trikojus to Temby. 24 February 1953.

41 R.W. Henderson and W.A. Rawlinson. 'Potentiometric and Other Studies on Preparations of Cytochrome *c* from Ox- and Horse-heart Muscle'. *Biochemical Journal*. v. 62 no. 1 (January 1956): 21–9.

42 Lindsay Sparrow. 'The Legacy of Pehr Edman'. *Australian Biochemist*. v. 33 no. 2 (August 2002): 16–21.

43 University of Melbourne staff file.

44 E. Minasian and N.A. Nicola. 'A Review of Cytokine Structures'. *Protein Sequences & Data Analysis*. v. 5 (1992): 57–64.

45 K. Sikaris, E. Minasian, S.J. Leach and R. Flegg. 'Computer Program Designed to Predict and Plot the Secondary Structure of Proteins'. *CABIOS*. v. 5 no. 4 (1989): 323.

46 D.A.D. Parry, E. Minasian and S.J. Leach. 'Conformational Homologies among Cytokines: Interleukins and Colony Stimulating Factors'. *Journal of Molecular Recognition*. Issue 1 (1988): 107–10; D.A.D. Parry, E. Minasian and S.J. Leach. 'Cytokine Conformations: Predictive Studies'. *Journal of Molecular Recognition*. Issue 4 (1991): 63–75.

47 UMA. Trikojus papers.

48 University of Melbourne Council. *Minutes*. 1 May 1967.

49 Aldo S. Bagnara and Lloyd R. Finch. 'Quantitative Extraction and Estimation of Intracellular Nucleoside Triphosphates of *Escherichia coli*'. *Analytical Biochemistry*. v. 45 no. 1 (January 1972): 24–34; Jack Maniloff (editor-in-chief). *Mycoplasmas: Molecular Biology and Pathogenesis*. Ronald N. McElhaney, Lloyd R. Finch and Joel B. Baseman (editors). Washington, DC: American Society for Microbiology, 1992.

Chapter 6—The Russell Grimwade School of Biochemistry

1 University of Melbourne Buildings Committee. *Minutes*. 15 February 1954.

2 Ibid.

3 UMA. Trikojus papers.

4 University of Melbourne Buildings Committee. *Minutes*. 18 August 1958.

5 V.M. Trikojus. 'Biochemistry and the University of Melbourne'. *Nature*. v. 191 no. 4795 (23 September 1961): 1238–40.

6 Ibid.

7 Ibid.

8 UMA. Trikojus papers.

9 Legge and Gibson. 'Victor Martin Trikojus, 1902–1985'.

10 Flesch. *Minding the Shop*. p. 149.

11 UMA. Trikojus papers.

12 Ibid.

13 'New C.S.I.R.O. Sugar Research Laboratory'. *Nature*. v. 187 no. 4735 (30 July 1960): 372.

14 UMA. Trikojus papers.

15 Ibid.

16 Ibid.

17 Ibid.

18 University of Melbourne Council. *Minutes*. February 1965.

19 Trikojus. 'Biochemistry and the University of Melbourne'.

20 Humphreys. *Trikojus*. p. 85.

21 M.R.Pawsey to R.E.H. Wettenhall. 16 February 1989. SAFETY\BIOSTAT.AB, File no: 11.280.5.

22 R.E.H. Wettenhall. Memorandum to All Members of Department. 20 December 1989.

23 Planning and Budgets Committee. *Minutes*. 12 April 2006.

24 Planning and Budgets Committee. *Minutes*. 5 December 2007.

25 Terrence Mulhern personal communication. 2013.

26 Personal conversation with Paul Gleeson. October 2013.

27 Staff file.

28 Ibid.

29 Ibid.

30 Ibid.

31 Personal correspondence from Bruce Livett. December 2013.

32 Flesch. *Minding the Shop*. p. 12; Christina Buckridge. 'Bracks Opens Uni's $100m Biotech R&D Flagship'. *UniNews*. v. 14 no. 11 (27 June – 11 July 2005).

Chapter 7—Stars Make Other Stars

1 http://www.med.monash.edu.au/biochem/news/lithgow-bio.html

2 P. Banks, W. Bartley and L.M. Birt. *The Biochemistry of the Tissues*. 2nd ed. London: Wiley, 1976. See also, for example, Alex M. Clarke and L.M. Birt. 'Evaluative Reviews in Universities: the Influence of Public Policies'. *Higher Education*. v. 11 no. 1b (January 1982): 1–26; Michael Birt. 'The Organisation of Tertiary Education in Australia: the Need for Re-arrangement'. *Journal of Tertiary Education Administration*. v. 7 no. 1 (1985): 22–34.

3 Personal conversation with Beverley Bencina 2013.

4 J. Shine and L. Dalgarno. 'The 3'-terminal Sequence of *Escherichia coli* 16S Ribosomal RNA: Complementarity to Nonsense Triplets and Ribosome

Binding Sites'. *Proceedings of the National Academy of Science USA*. v. 71 (1974): 1342–6.

5 L.W. Nichol, W.J.H. Jackson and D.J. Winzor. 'A Theoretical Study of the Binding of Small Molecules to a Polymerizing Protein System. A Model for Allosteric Effects'. *Biochemistry*. v. 6 no. 8 (1967): 2449–56; G.H. McKenzie, W.H. Sawyer and L.W. Nichol. 'The Molecular Weight and Stability of Concanavalin A'. *Biochimica et Biophysica Acta (BBA) – Protein Structure*. v. 263 no. 2 (15 April 1972): 283–93.

6 A.E. Clarke and B.A. Stone. 'β-1, 3-Glucan Hydrolases from the Grape Vine (*vitis Vinifera*) and Other Plants'. *Phytochemistry*. v. 1 no. 3 (September 1962): 175–88.

7 Bruce A. Stone and Adrienne Clarke. *Chemistry and Biology of (1-3)-Beta-Glucans*. Melbourne: La Trobe University Press, 1992.

8 For a fuller obituary tribute, see Nick Hoogenraad. 'In Memoriam: Bruce Arthur Stone (1928–2008)'. *Australian Biochemist*. v. 39 no. 3 (December 2008): 28–9.

9 Anthony B. Blakeney, Philip J. Harris, Robert J. Henry and Bruce A. Stone. 'A Simple and Rapid Preparation of Alditol Acetates for Monosaccharide Analysis'. *Carbohydrate Research*. v. 113 no. 2 (1 March 1983): 291–9.

10 http://www.eoas.info/biogs/P002212b.htm

11 Adrienne Clarke, Paul Gleeson, Susan Harrison and R. Bruce Knox. 'Pollen Stigma Interactions: Identification and Characterization of Surface Components with Recognition Potential.' *Proceedings of the National Academy of Sciences*. v. 76 no. 7 (1979): 3358–62.

12 http://sciencearchive.org.au/scientists/interviews/h/nhoogenraad.html

13 Ibid.

14 N.J. Hoogenraad, F.J.R. Hird, I. Holmes and Nancy F. Millis. 'Bacteriophages in Rumen Contents of Sheep'. *Journal of General Virology*. v. 1 no. 4 (October 1967): 575–6.

15 A.W. Burgess and D. Metcalf. 'Characterization of a Serum Factor Stimulating the Differentiation of Myelomonocytic Leukemic Cells'. *International Journal of Cancer*. v. 26 (1980): 647–54; T.P. Garrett, A.W. Burgess, H.K. Gan, R.B. Luwor, G. Cartwright, F. Walker, S.G. Orchard et al. 'Antibodies Specifically Targeting a Locally Misfolded Region of Tumor Associated EGFR'. *Proceedings of the National Academy of Sciences*. v. 106 (2009): 5082–7; L.G. Sparrow, D. Metcalf, M.W. Hunkapiller, L.E. Hood and A.W. Burgess. 'Purification and Partial Amino Acid Sequence of Asialo Murine Granulocyte-Macrophage Colony Stimulating Factor'. *Proceedings of the National Academy of Sciences*. v. 82 (1985): 292–6; F. Walker, H.-H. Zhang, A. Odorizzi and A.W. Burgess. 'LGR5 Is a Negative Regulator of Tumourigenicity, Antagonizes Wnt Signalling and Regulates Cell Adhesion in Colorectal Cancer Cell Lines'. *PloS one*. v. 6 (2011): e22733.

16 R.A. Burton, S.M. Wilson, M. Hrmova, A.J. Harvey, N.J. Shirley, A. Medhurst, B.A. Stone, E.J. Newbigin, A. Bacic and G.B. Fincher. 'Cellulose Synthase-like CslF Genes Mediate the Synthesis of Cell Wall (1,3;1,4)-β-Glucans'. *Science*. v. 311

(2006): 1940–2; A.K. Jacobs, V. Lipka, R.A. Burton, R. Panstruga, N. Strizhov, P. Schulze-Lefert and G.B. Fincher. 'An *Arabidopsis thaliana* Callose Synthase, GSL5, Is Required for Wound and Papillary Callose Formation'. *Plant Cell*. v. 15 (2003): 2503–13.

17   Matthew Anthony Perugini. 'Characterization of Human Apolipoprotein E3 and E4 Isoforms'. Thesis (PhD). University of Melbourne, School of Biochemistry and Molecular Biology, 2002.

18   B.R. Burgess, R.C.J. Dobson, M.F. Bailey, S.C. Atkinson, M.D.W. Griffin, G.B. Jameson, M.W. Parker, J.A. Gerrard and M.A. Perugini. 'Structure and Evolution of a Novel Dimeric Enzyme from a Clinically Important Bacterial Pathogen'. *Journal of Biological Chemistry*. v. 41 no. 283 (2008): 27598–603.

19   Barrie Davidson, Elizabeth H. Blackburn and Theo Dopheide. 'Chorismate Mutase-Prephenate Dehydratase from *Escheria coli* K-12'. *Journal of Biological Chemistry*. v. 247 no. 14 (25 July 1972): 441–6.

20   For example, Elizabeth H. Blackburn and F.J.R. Hird. 'Metabolism of Glutamine and Glutamate by Rat Liver Mitochondria'. *Archives of Biochemistry and Biophysics*. v. 152 no. 1 (September 1972): 258–64.

21   Letter from Hird to Mrs M. Blackburn, 10 March 1983, quoted in Catherine Brady. *Elizabeth Blackburn and the Story of Telomeres: Deciphering the Ends of DNA*. Cambridge, Mass.: MIT Press, 2007. p. 20.

22   Jane Carey. '"What's a Nice Girl Like You Doing with a Nobel Prize?" Elizabeth Blackburn, "Australia's" First Woman Nobel Laureate and Women's Scientific Leadership'. http://www.womenaustralia.info/leaders/sti/pdfs/19_Carey.pdf

23   Elizabeth H. Blackburn and Elissa S. Epel. 'Too Toxic to Ignore'. *Nature*. v. 490 (11 October 2012): 169–71.

24   Rebecca Scott. 'Elizabeth Blackburn School of Sciences Launched'. *Voice*. v. 10 no. 4 (14 April – 11 May 2014): 6.

25   'Another Win for Women'. *Age*. 9 November 1971: 12.

26   Mary-Jane Gething oral history interview. Cold Spring Harbor Laboratory Library. 16 January 2003. http://library.cshl.edu/oralhistory/category/cshl/

27   Mary-Jane Gething and Joseph Sambrook. 'Protein Folding in the Cell'. *Nature*. v. 355 no. 6355 (2 January 1992): 33–45.

28   M. Sweet. 'The Profile: Kerin O'Dea'. *Australian Doctor*. July 2000: 27–9.

29   Kerin O'Dea. 'Marked Improvement in Carbohydrate and Lipid Metabolism in Diabetic Australian Aborigines after Temporary Reversion to Traditional Lifestyle'. *Diabetes*. v. 33 no. 6 (June 1984): 596–603.

30   Ibid. Summary.

31   http://outbackstores.com.au/

32   Marno C. Ryan, Catherine Itsiopoulos, Tania Thodis, Glenn Ward, Nicholas Trost, Sophie Hofferberth, Kerin O'Dea, Paul V. Desmond, Nathan A. Johnson and Andrew M. Wilson. 'The Mediterranean Diet Improves Hepatic Steatosis and Insulin Sensitivity in Individuals with Non-Alcoholic Fatty Liver Disease'. *Journal of Hepatology*. v. 59 no. 1 (July 2013): 138–43.

33   http://www.florigene.com/?country=Australia

34   Elaine Mulcahy. 'Leading Dental Research Wins Uni Scientist Victoria Prize'. *UniNews.* v. 14 no. 15 (22 August – 5 September 2005).

35   Colin E. Moore. 'Herbert Frank Shorney MD, FRCS (1878 – 2 May 1933)'. *Australian Journal of Ophthalmology.* v. 12 (1984): 289–91.

36   Robert C. Augusteyn and Jane F. Koretz. 'Hypothesis: a Possible Structure For α-crystallin'. *FEBS Letter.* v. 222 no. 1 (28 September 1987): 1–5.

37   Barry L. Cole. 'Robert Augusteyn: Director of the National Vision Research Institute of Australia 1991 to 2001'. *Clinical and Experimental Optometry.* v. 85 no. 2 (March 2002): 107–10.

38   See, for example, Robert C. Augusteyn, Derek Nankivil, Ashik Mohamed, Bianca Maceo, Faradia Pierre and Jean-Marie Parel. 'Human Ocular Biometry'. *Experimental Eye Research.* v. 102 (2012): 70–5; Robert C. Augusteyn, Ashik Mohamed and Jean-Marie Parel. 'Lens Thickness Growth in Humans'. *Clinical and Experimental Ophthalmology.* v. 41 no. 6 (August 2013): 616–17.

39   R.C. Augusteyn, H.B. Collin and K.M Rogers. *The Eye.* Montreal: Eden Press, 1979, p. 220.

40   R.I. Christopherson. 'Enzyme Inhibitors Show Promise as Anti-cancer Drugs'. *MU Research.* v. 2 no. 2 (March 1986): 3–4.

41   Seth L. Masters, Geoffrey J. Howlett and Richard N. Pau. 'The Molybdate Binding Protein Mop from *Haemophilus influenzae*: Biochemical and Thermodynamic Characterisation'. *Archives of Biochemistry and Biophysics.* v. 439 (2005): 105–12; Werner Klipp, Bernd Masepohl, John R. Gallon and William E. Newton (editors). *Genetics and Regulation of Nitrogen Fixation in Free-Living Bacteria.* New York: Kluwer, 2005.

42   David Ebert. 'Great Expectations'. *Australian Biochemist.* v. 35 no. 3 (December 2004): 26–9.

43   'Medical Students Vote to Identify Teaching Excellence'. *UniNews.* v. 12 no. 6 (21 April – 5 May 2003). http://archive.uninews.unimelb.edu.au/news/550/

44   Ebert. 'Great Expectations'.

45   Robin Fredric Anders. 'Aspects of the Growth and Metabolism of Group N Streptococci'. Thesis (PhD). University of Melbourne, 1967.

46   Melissa Anne Brown. 'The GM-CSF Receptor and Human Leukaemia'. Thesis (PhD). University of Melbourne, 1993.

Chapter 8—'The Academics Couldn't Function without Them'

1   University of Melbourne staff file.

2   Ibid.

3   University of Melbourne Staff and Establishments Committee. *Minutes.* 22 February 1954.

4   http://www.ww2roll.gov.au/Veteran.aspx?ServiceId=A&VeteranId=504719 and University of Melbourne staff file.

5   UMA. Trikojus papers. Letter of application to Gallacher. 5 March 1963.

6   Ibid.

7   UMA. Trikojus papers.

8   Ibid.

9   Personal communication.

10 Kevin M. Downard and John R. de Laeter. 'A History of Mass Spectrometry in Australia'. *Journal of Mass Spectrometry*. v. 40 (2005): 1123–39.

11 Text kindly provided by John Morley.

12 University of Melbourne staff file.

13 Ibid.

14 Mark VK3PI. 'Vale Robert (Bob) Goullet, VK3BU'. *WANSARC News*. v. 40 no. 4 (May 2009): 2.

15 University of Melbourne staff file.

16 Ibid.

17 University of Melbourne. *Staff News*. 17 August 1970: 6.

18 University of Melbourne staff file.

19 Flesch. *Minding the Shop*. pp. 17–30 list some of the longest-serving porters.

20 See Flesch. *Life's Logic*. pp. 206–7 for the Medical Building.

21 Ibid. pp. 30–5.

22 Flesch. *Minding the Shop*. pp. 33–5 provide more detail.

23 University of Melbourne staff file.

24 SafetyMAP (Safety Management Achievement Program) was an audit development standard developed by the Victorian WorkCover Authority.

25 Paul Gleeson. 'Farewell to the Russell Grimwade School of Biochemistry Building, University of Melbourne'. *Australian Biochemist*. v. 39 no. 2 (August 2008): 28–31.

26 B.J. Smith and M.M. Gagnon. 'MFO Induction of Three Australian Fish Species'. *Environmental Toxicology*. v. 15 (2000): 1–7.

27 University of Melbourne staff file.

28 Ibid. See also Paul Cagliore, D. James, B. St John, Lindsay J. Rayner and John W. Legge. 'Peroxidase Levels in Food: Relevance to Colorectal Cancer Screening'. *American Journal of Clinical Nutrition*. v. 35 (June 1982): 1487–9.

29 Ibid.

Chapter 9—On the Move Again

1   For example, A. Mitchell, I.L. Sin and L.R. Finch. 'Enzymes of Purine Metabolism in *Mycoplasma mycoides subsp mycoides*'. *Journal of Bacteriology*. v. 134 (1978): 706–12.

2   For example, D. Mathai, N. Fidge, M. Tozuka and A. Mitchell. 'Regulation of Hepatic High Density Lipoprotein Binding Proteins after Administration of Simvastatin and Cholestyramine to Rats'. *Arteriosclerosis*. v. 10 (1990): 1045–50.

3   Patricia Kahn, Lars Frykberg, Claire Brady, Irene Stanley, Hartmut Beug, Björn Vennström and Thomas Graf. 'v-erbA Cooperates with Sarcoma Oncogenes in Leukemic Cell Transformation'. *Cell*. v. 45 (May 1986): 349–56.

4 'Teaching Good Value'. *Voice*. v. 5 no. 1 (13 April – 10 May 2009). http://archive. uninews.unimelb.edu.au/news/5772/

5 G.R. Parslow. *The 'Q' Instruction Package*. Cambridge: Elsevier-Biosoft Publishers, 1992. First edition 1987. Version 3.0 released 1992.

6 For example, Graham R. Parslow. 'Multimedia in Biochemistry and Molecular Biology Education, Commentary: Massive Open Online Courses'. *Biochemistry and Molecular Biology Education*. v. 41 no. 4 (August 2013): 278–9.

7 Leon Helfenbaum. 'Molecular and Genetic Investigations into the Expression of Yeast Mitochondrial ATP Synthase Subunit 9'. Thesis (PhD). Monash University, 1995.

8 B. Pal, N.C. Chan, L. Helfenbaum, K. Tan, W.P. Tansey and M.-J. Gething. 'SCF Cdc4-Mediated Degradation of the Hac1p Transcription Factor Regulates the Unfolded Protein Response in *Saccharomyces Cerevisiae*'. *Molecular Biology of the Cell*. v. 18 (2007): 426–40.

9 For example: A. Willems-Jones et al. 'High Grade Prostatic Intraepithelial Neoplasia Does Not Display Loss of Heterozygosity at the Mutation Locus in BRCA2 Mutation Carriers with Aggressive Prostate Cancer. *BJU International*. Epub. 4 Ocober 2012; H. Thorne, A. Willems-Jones et al. 'Decreased Prostate Cancer-specific Survival of Men with *BRCA2* Mutations from Multiple Breast Cancer Families'. *Cancer Prevention Research*. v. 4 (2011): 1002–10.

10 Personal correspondence. 2013.

11 R.D. Mills, J. Trewhella, T.W. Qiu, T. Ryan, T. Hanley, R.B. Knott, T. Lithgow and T.D. Mulhern. 'Domain Organization of the Monomeric Form of the Tom70 Mitochondrial Import Receptor'. *Journal of Molecular Biology*. v. 338 (2009): 1043–58.

12 University of Melbourne advertisement: http://uom.clients.pageup.com.au/jobDetails.asp?sJobIDs=851221&stp=AW&sLanguage=en

13 Thomas E. Skinner and M. Robin Bendall. 'A Vector Model of Adiabatic Decoupling'. *Journal of Magnetic Resonance*. v. 134 (1998): 315–30; M. Robin Bendall and Thomas E. Skinner. 'Calibration of STUD+ Parameters to Achieve Optimally Efficient Broadband Adiabatic Decoupling in a Single Transient'. *Journal of Magnetic Resonance*. v. 134 (1998): 331–49.

14 See, for example, Anthony W. Purcell, James McCluskey and Jamie Rossjohn. 'More than One Reason to Rethink the Use of Peptides In Vaccine Design'. *Nature Reviews Drug Discovery*. v. 6 (May 2007): 404–14.

15 P.T. Illing, J.P. Vivian, N.L. Dudek, L. Kostenko, Z. Chen, M. Bharadwaj, J.J. Miles, L. Kjer-Nielsen, S. Gras, N.A. Williamson, S.R. Burrows, A.W. Purcell*, J. Rossjohn* and J. McCluskey*. 'Immune Self-reactivity Triggered by Drug-modified Human Leukocyte Antigen-peptide Repertoire'. *Nature*. v. 486 (2012): 554–8. (* co-corresponding authors)

16 The Australian Society for Biochemistry and Molecular Biology's Boomerang Award recognises the research of an Australian biochemist or molecular biologist based outside Australia with no more than five years' postdoctoral

experience. The Young Investigator Award of the Annual Lorne Conference on Protein Structure and Function recognises a protein biochemist of no more than five years' postdoctoral experience.

17   Y.M. Ramdzan, S. Polling, C.P.Z Chia, I.H.W. Ng, A.R. Ormsby, N.P. Croft, A.W. Purcell, M.A. Bogoyevitch, D.C.H. Ng, P.A. Gleeson, P.A. Hatters and D.M. Hatters. 'Tracking Protein Aggregation and Mislocalization in Cells with Flow Cytometry'. *Nature Methods*. v. 9 (2012): 467–70.

18   David Guest and Bruce Grant. 'The Complex Action of Phosphonates as Antifungal Agents'. *Biological Reviews*. v. 66 no. 2 (May 1991): 159–87.

19   '$10 000 Grant for Cinnamon Fungus Research'. *University Gazette*. March 1981: 14. Also, for example: David M. Cahill, Gretna M. Weste and Bruce R. Grant. 'Changes in Cytokinin Concentrations in Xylem Extrudate Following Infection of *Eucalyptus marginata* Donn ex Sm with *Phytophthora cinnamomi* Rands 1'. *Plant Physiology*. v. 81 no. 4 (August 1986): 1103–9.

20   University of Melbourne staff file. See, for example, Helen R. Irving, Julia M. Griffith and Bruce R. Grant. 'Calcium Efflux Associated with Encystment of Zoospores'. *Cell Calcium*. v. 5 no. 5 (October 1984): 487–500.

21   R. Bruce Knox and Michael Tuohy. 'Pollen, Plants and People: a Review of Pollen Aerobiology in Southern Australia'. In *Proceedings of the Sixth Australian Weeds Conference, Broadbeach International Hotel, City of Gold Coast, Queensland: 13–18 September 1981*. Edited by B.J. Wilson and J.T. Swarbrick. Queensland: Queensland Weed Society for the Council of Australian Weed Science Societies, 1981. pp. 125–42.

22   University of Melbourne staff file.

23   Ibid.

24   Ibid.

25   Robert Hannaford (1994–). *University Council 2003*. 2006–07 oil on canvas. The University of Melbourne Art Collection 2007.0003.000.000.

26   http://www.unimelb.edu.au/unisec/calendar/honcausa/citation/anning.pdf

27   http://www.nhmrc.gov.au/guidelines/publications/ea28

28   Personal correspondence. 2013.

29   For example: R. Aebersold, G.D. Pipes, R.E. Wettenhall, H. Nika and L.E. Hood. 'Covalent Attachment of Peptides for High Sensitivity Solid-phase Sequence Analysis'. *Analytical Biochemistry*. v. 187 no. 1 (15 May 1990): 56–65; R.E. Wettenhall, R.H. Aebersold and L.E. Hood. 'Solid-phase Sequencing of 32P-Labeled Phosphopeptides at Picomole and Subpicomole Levels'. *Methods in Enzymology*. v. 201 (1991): 186–99.

30   D.M. Robertson, L.M. Foulds, L. Leversha, F.J. Morgan, M.T.W. Hearn, H.G. Burger, R.E.H. Wettenhall and D.M. de Kretser. 'Isolation of Inhibin from Bovine Follicular Fluid'. *Biochemical and Biophysical Research Communications*. v. 126 no. 1 (16 January 1985): 220–6.

31   University of Melbourne Council. *Minutes*. 3 July 2000.

Chapter 10—The New World

1 University of Melbourne. *Annual Report 2003*. p. 21.

2 University of Melbourne. *Annual Report 2001*. p. 3.

3 http://www.bio21.unimelb.edu.au/about-bio21

4 Kenwyn R. Gayler and Geoffrey E. Sykes. 'Effects of Nutritional Stress on the Storage Proteins of Soybeans'. *Plant Physiology*. v. 78 no. 3 (July 1985): 582–5.

5 Pallavi Bhosle and Sawarkar Vaibhav. 'Conotoxins: Possible Therapeutic Measure for Huntingtons Disease'. *Journal of Neurological Disorders*. v. 1 no. 3 (2013): 129–32.

6 Jason Major. 'Snail Venom May Herald a New Era in the Treatment of Chronic Pain'. *UniNews*. 29 July – 12 August 2002: 1, 7.

7 D.W. Sandall, N. Satkunanathan, D.A. Keays, M.A. Polidano, X. Liping, V.Pham, J.G. Down, Z. Khalil, B.G. Livett and K.R. Gayler. 'A Novel α-Conotoxin Identified by Gene Sequencing Is Active in Suppressing the Vascular Response to Selective Stimulation of Sensory Nerves in Vivo'. *Biochemistry*. v. 42 no. 22 (2003): 6904–11.

8 http://grimwade.biochem.unimelb.edu.au/cone

9 H C. Cheng, B.E. Kemp, R.B. Pearson, A.J. Smith, L. Misconi, S.M. Van Patten and D.A. Walsh. 'A Potent Synthetic Peptide Inhibitor of the cAMP-dependent Protein Kinase'. *Journal of Biological Chemistry*. v. 261 (1986): 989–92.

10 M.I. Hossain, C.L. Roulston, M.A. Kamaruddin, P.W.Y. Chu, Dominic C.H. Ng, Gregory J. Dusting, Jeffrey D. Bjorge, Nicholas A. Williamson, Donald J. Fujita, Steve N. Cheung, Tung O. Chan, Andrew F. Hill and Heung-Chin Cheng. 'A Truncated Fragment of Src Protein Kinase Generated by Calpain-Mediated Cleavage Is a Mediator of Neuronal Death in Excitotoxicity'. *Journal of Biological Chemistry*. v. 288 (2013): 9696–709.

11 P.R. Gooley, B.A. Johnson, A.I. Marcy, G.C. Cuca, S.P. Salowe, W.K. Hagmann, C.K. Esser and J.P. Springer. 'Secondary Structure and Zinc Ligation of Human Recombinant Short-form Stromelysin by Multidimensional Heteronuclear NMR'. *Biochemistry*. v. 32 (1993): 13098–108; P.R. Gooley, J.F. O'Connell, A.I Marcy, G.C. Cuca, S.P. Salowe, B.L. Bush, J.D. Hermes, W.K. Hagmann, C.K. Esser, J.P. Springer and B.A. Johnson. 'The NMR Structure of the Inhibited Catalytic Domain of Human Stromelysin-1'. *Nature Structural Biology*. v. 1 (1994): 111–18; P.R. Gooley, J.F. O'Connell, A.I. Marcy, G.C. Cuca, W.K. Hagmann, C.G. Caldwell, M.G. Axel and J.W. Becker. 'Comparison of the Structure of Human Recombinant Short-form Stromelysin by Multidimensional Heteronuclear NMR and X-ray Crystallography'. *Journal of Biomolecular NMR*. v. 7 (1996): 8–28.

12 Ian R. Van Driel, Andrew F. Wilks, Geoffrey A. Pietersz and James W. Goding. 'Murine Plasma Cell Membrane Antigen PC-1: Molecular Cloning of cDNA and Analysis of Expression'. *Proceedings of the National Academy of Sciences of the USA*. v. 82 (December 1985): 8619–23.

13 Nhung Nguyen, Louise M. Judd, Anastasia Kalantzis, Belinda Whittle, Andrew S. Giraud and Ian R. van Driel. 'Random Mutagenesis of the Mouse Genome: a Strategy for Discovering Gene Function and the Molecular Basis of Disease'. *American Journal of Physiology – Gastrointestinal and Liver Physiology*. v. 300 no. 1 (January 2011): G1–G11.

14 J. Collinge, K.C.L. Sidle, J. Meads, J. Ironside and A.F. Hill. 'Molecular Analysis of Prion Strain Variation and the Aetiology of "New Variant" CJD'. *Nature*. v. 383 (1996): 685–90.

15 Denis Crane and Marie Bogoyevitch. 'Great Expectations'. *Australian Biochemist*. v. 38 no. 1 (April 2007): 28–30.

16 Ibid. p. 30.

17 M.A. Bogoyevitch and B. Kobe. 'Uses for JNK: the Many and Varied Substrates of the c-Jun N-terminal Kinases'. *Microbiology and Molecular Biology Reviews*. v. 70 no 4 (2006): 1061–95.

18 Y.Q. Wong, K.J. Binger, G.J. Howlett and M.D.W. Griffin. 'Methionine Oxidation Induces Amyloid Fibril Formation by Full-length Apolipoprotein A-I'. *Proceedings of the National Academy of Sciences U.S.A.* v. 107 (2010): 1977–82.

19 Nectarios Klonis, Stanley C. Xie, James M. McCaw, Maria P. Crespo, Sophie G. Zaloumis, Julie A. Simpson and Leann Tilley. 'Altered Temporal Response of Malaria Parasites Determines Differential Sensitivity to Artemisinin'. *Proceedings of the National Academy of Sciences of the United States of America*. v. 110 no. 13 (2013): 5157–62.

20 Leann Tilley. 'Audrey Cahn: a Nonagenarian Scientist Remembers the Early Days'. *Wisenet Journal* no. 49 (1998): 10–13; Leann Tilley and Bruce Stone. 'A Scientist Ahead of Her Times: Audrey Josephine Cahn: Nutritionist, Artist, 17.10.1905-1.4.2008'. *Age*. 12 May 2008.

21 José A. Villadangos and Petra Schnorrer. 'Intrinsic and Cooperative Antigen Presenting Functions of Dendritic-cell Subsets *in vivo*.' *Nature Reviews Immunology*. v. 7 no. 7 (July 2007): 543–55; Linda M. Wakim, Nishma Gupta, Justine D. Mintern and José A. Villadangos. 'Enhanced Survival of Lung Tissue-resident Memory CD8+ T cells during Infection with Influenza Virus Due to Selective Expression of IFITM3'. *Nature Immunology*. v. 14 (27 January 2013): 238–45.

22 Kathryn Elizabeth Holt. 'Genomic Variation and Evolution of *Salmonella enterica* serovars Typhi and Paratyphi A'. Thesis (PhD). University of Cambridge, 2009.

Chapter 11—A Bridging Science

1 E.M. Hume. 'Obituary: Charles James Martin, Kt, C.M.G., F.R.C.P., D.Sc., F.R.S. (9 January 1866 – 15 February 1955)'. *British Journal of Nutrition*. v. 10 no. 1 (1956): 1–7.

2 The Bio21 Institute. 'Researchers to Benefit from a Share of More than $8.1 Million Awarded in Latest Round of NHMRC Funding'. 29 October 13.

http://www.bio21.unimelb.edu.au/news/researchers-to-benefit-from-a-share-of-more-than-8-1-million-awa

3   H. Safavi-Hemami, D.G. Gorasia, P. Veith, E. Reynolds, P. Bandyopadhyay, B.M. Olivera and A.W. Purcell. 'The Regional Proteome of the Venom Gland of the Fish-hunting Cone Snail *Conus geographus* Reveals Distinct Compartimentalisation of Proteins'. *Molecular and Cellular Proteomics*. Epub 29 January 2014.

4   Shayne Anthony Bellingham. 'Copper Homeostasis and the Alzheimer's Disease Amyloid Precursor Protein'. Thesis (PhD). University of Melbourne, 2005.

5   Joe Fennessy. 'New Partnerships Support Indigenous Health'. *Voice*. v. 10 no. 1 (13 January – 9 February 2014).

6   The Melbourne Newsroom. '"Mad Cow" Blood Test Now on the Horizon'. 12 September 2012. http://newsroom.melbourne.edu/news/n-901

7   Megan K. Dearnley, Jeffrey A. Yeoman, Eric Hanssen, Shannon Kenny, Lynne Turnbull, Cynthia B. Whitchurch, Leann Tilley and Matthew W.A. Dixon. 'Origin, Composition, Organization and Function of the Inner Membrane Complex of *Plasmodium falciparum* Gametocytes'. *Journal of Cell Science*. v. 125 (2012): 2053–63.

8   Nerissa Hannink. 'Malaria Goes Bananas before Reproduction'. *Voice*. v. 8 no. 3 (12 March – 8 April 2012).

9   P.Z.C. Chia. 'Dissecting Retrograde Transport Pathways in Development and Disease'. Thesis (PhD). University of Melbourne, 2011.

10 P.Z.C. Chia, W.H. Toh, R.A. Sharples, I. Gasnereau, A.F. Hill and P.A. Gleeson. 'Intracellular Itinerary of Internalised Beta-secretase BACE1, and Its Potential Impact on Amyloid Precursor Protein Processing'. *Traffic*. v. 14 no. 9 (2013): 997–1013.

Acknowledgements

FIRST AND FOREMOST I thank the Head of the Department of Bio-chemistry and Molecular Biology, Paul Gleeson, for commissioning my work. I also thank every member of the department staff, past and present, who took an interest, providing anecdotes, information, photographs and reminiscences, even if these could not always be printed. Almost every living person mentioned in the text helped in its production. It is perhaps invidious to single out individuals, but I especially acknowledge the help of Lindsay Rayner and Beverley Bencina in all aspects of my work. Lynn Tran and Brent Smith gave special help with the photographs. Brett Drummond generously shared an office with me for two years and Jim Dang was unfailingly kind about my incompetence with computers.

From the University Library and University Archives, I especially thank Jane Beattie, Katie Wood, Leanne McCredden, Jim Berryman, Lea McRae and Susan Millard. Linda Notley from the State Library of Victoria and the staff of the ACT Planning and Land Authority were, as ever, invaluable. As he has been so many times before, Kevin Whitton was the source of a wonderful amount of genealogical detail. Suzanne Blom from the Advancement Office, Anna Frith and Sheridan Nanscawen from Human Resources, Leanne Dyson, Kathryn Dan and Steve Halliwell from the University Secretary's office and Jacquie Munro-Smith and Liz Moon from Bio21 were all helpful and informative. From the Faculty of Science,

Robyn Trethowan provided useful information on the Grimwade Prize. Marc Cheeng and Jade Germantis from Property and Campus Services provided information on buildings and plans. Benjamin Thomas, co-author of a forthcoming book on the Grimwade Bequest, most generously allowed me to see drafts of his work. I am grateful to the Rothera family for responding positively to my requests for yet more information about the university's first lecturer in Biochemistry. Cathryn Smith of Melbourne University Publishing was both kind and patient with me and Paul Smitz did a great job of copy-editing and coping with my disquiet about some aspects of the house style. Judith Tuck, as she has done so often in the past, demonstrated an eagle eye for grammatical mistakes and typographical errors as well as suggesting some of the funnier possible titles for this book.

From La Trobe University I thank Nick Hoogenraad, Ian Potter and John Hill. I also thank John Hunt, Ann Eames, Matt Perugini, Neil Griffiths, the McEachern family, Jackie Camilleri-Zahra, John Morley, Evelyn Parkhill, Amber Willems-Jones, Graham Parslow and Jaynie Anderson for personal photographs and information. Carla Flores of CSIRO greatly facilitated use of the photograph of Syd Leach. Liana Friedman of ASBMB and Chu Kong Liew allowed use of the photograph of David Ebert and they and Stephen Collins that of Fred Collins. Lynette Mahoney and Karen Rogers were of great assistance with the images held by Monash University.

From the University of Melbourne at large I thank Peter McPhee for allowing me to bash his ear on the bus to the university and for providing me with a different project to take my mind off Biochemistry occasionally, and the late great Valda McRae for her friendship and willingness to share her unequalled fund of knowledge about the University of Melbourne.

Any errors in the finished work are my responsibility alone.

Photography Credits

Chemical Physiology Classroom, University of Melbourne Archives, p. *xviii*.

George Britton Halford, courtesy of Medical History Museum Collection, p. 2.

John Macadam, courtesy of State Library of Victoria, p. 2.

John Drummond Kirkland, courtesy of Medical History Museum Collection, p. 5.

William Alexander Osborne, courtesy of Medical History Museum Collection, p. 5.

Arthur Rothera, courtesy of Derek Rothera, p. 7.

Plaque, courtesy of Richard Kirsner, p. 10.

Lilias Maxwell (nee Jackson), courtesy of John Hunt, p. 11.

Ivan Maxwell, courtesy of John Hunt, p. 12.

William John Young, courtesy of Medical History Museum Collection, p. 15.

Sir Robert Menzies and Victor Trikojus, Trikojus papers, University of Melbourne Archives, p. 16.

Viktor Martin Trikojus, University of Melbourne Archives, p. 20.

Francis Gordon Lennox, courtesy of Monash University Archives (IN 2106), p. 27.

The Nicholas Nutrition Laboratory, courtesy of Media & Publications Unit, p. 32.

Jean Millis Jackson, courtesy of Nancy Millis, p. 35.

Audrey Cahn, courtesy of Leann Tilley, p. 39.

Muriel Crabtree, courtesy of University College, p. 41.

Mary McQuillan, Staff file, p. 44.

The Department of Biochemistry in 1948, Trikojus papers, University of Melbourne Archives, p. 46.

William Rawlinson, courtesy of Ann Eames, p. 49.

Peter Springell, courtesy of National Archives of Australia (NAA: A1200, L53159), p. 55.

Francis Hird, Trikojus papers, courtesy of University of Melbourne Archives, p. 57.

Jack Legge, courtesy of Australian War Memorial (P05253.001), p. 60.

Robert Morton, courtesy of Australian Academy of Science, p. 63.

Joseph Bornstein, courtesy of Monash University Archives (IN 289), p. 65.

Max Marginson, courtesy of University of Melbourne Archives, p. 67.

Pamela Todd, courtesy of St Hilda's College, p. 69.

Archibald McEachern, courtesy of Gregory McEachern, p. 72.

Joyce Calvert, courtesy of John Morley, p. 73.

The Department of Biochemistry in 1978, courtesy of Media & Publications Unit, p. 74.

Russell Grimwade, Grimwade Collection, University of Melbourne Archives, p. 78.

Evelyn Parkhill, courtesy of John Morley, p. 82.

Peter Hall, Staff file, p. 83.

Syd Leach, © CSIRO, p. 84.

Frederick Collins, courtesy of Stephen Collins, p. 86.

Gerhard Schreiber, courtesy of John Morley, p. 88.

Barry Davidson and Richard Pau, courtesy of Amber Willems-Jones, p. 90.

William Sawyer and LeRoy Henderson, courtesy of Media & Publications Unit, p. 92.

Robert Henderson, courtesy of Janet Rowswarne, p. 94.

Elizabeth Minasian and Agnes Henschen, courtesy of Australian Society for Biochemistry and Molecular Biology, reproduced with permission of St Vincent's Institute of Medical Research, p. 96.

Lloyd Finch, courtesy of Lloyd Finch, p. 292.

Russell Grimwade Building plan, courtesy of Property & Campus Services, University of Melbourne, p. 100.

Mab Grimwade, Grimwade Collection, University of Melbourne Archives, p. 105.

Russell Grimwade Building, Grimwade Collection, University of Melbourne Archives, pp. 109, 119, 123.

Pip Pattison, Jane Gunn, Ingrid Scheffer, Judith Whitworth, Elizabeth Blackburn, Elizabeth Hartland, Leann Tilley and Ruth Bishop, by Peter Casamento, p. 126.

Trevor Lithgow, courtesy of Monash University Archives, p. 128.

Michael Birt and Ray Martin, by Tony Miller, courtesy of Monash University Archives (IN 2691), p. 130.

Lynn Dalgarno, courtesy of Lynn Dalgarno, p. 132.

Lawrence Walter Nichol, University of Melbourne Archives, p. 134.

Bruce Stone, courtesy of La Trobe University, p. 135.

Adrienne Clarke, University of Melbourne Archives, p. 136.

Suzanne Cory, courtesy of Suzanne Cory, p. 137.

Nick Hoogenraad, courtesy of Nick Hoogenraad, p. 139.

Antony Burgess, courtesy of Tony Burgess, p. 140.

Geoffrey Fincher, courtesy of Waite Institute, p. 141.

Matthew Perugini, courtesy of Matt Perugini, p. 142.

Elizabeth Blackburn, courtesy of Creative Commons, p. 144.

Mary-Jane Gething, University of Melbourne Archives, p. 145.

Kerin O'Dea, courtesy of Kerin O'Dea, p. 147.

Edwina Cornish, courtesy of Monash University, p. 148.

Robert Augusteyn, University of Melbourne *Staff News* 1979, p. 151.

Richard Christopherson, courtesy of Richard Christopherson, p. 153.

David Ebert and Craig Clements, courtesy of David Ebert, p. 156.

Melissa Brown, courtesy of University of Queensland, p. 160.

Gleeson Laboratory, courtesy of John Morley, p. 162.

John Chippindall, courtesy of La Trobe University, p. 168.

Francis Baker, courtesy of John Morley, p. 170.

Neil Griffiths, courtesy of Neil Griffiths, p. 172.

Robert Goullet, Albert Fairchild, John Morley and Lindsay Rayner, courtesy of John Morley, p. 175.

Graeme Strange, Staff file, p. 176.

John Morley, courtesy of John Morley, p. 180.

Jacqueline Camilleri, courtesy of Jackie Camilleri, p. 182.

Beverley Bencina, courtesy of Beverley Bencina, p. 185.

Brent Smith, by Lynn Tran, p. 185.

Lynn Tran, by Brent Smith, p. 186.

Lindsay Rayner, John Morley and Jackie Camilleri, courtesy of John Morley, p. 188.

Teaching Laboratory, courtesy of Paul Gleeson, p. 190.

Alana Mitchell, by Brent Smith, p. 192.

Irene Stanley, courtesy of Irene Stanley, p. 193.

Graham Parslow, by Irene Stanley, p. 196.

Leon Helfenbaum, by Irene Stanley, p. 197.

Amber Willems-Jones, courtesy of Amber Willems-Jones, p. 198.

Terrence Mulhern, courtesy of Terry Mulhern, p. 200.

Phillip Dickson, courtesy of Phillip Dickson, p. 203.

Anthony Purcell, courtesy of Tony Purcell, p. 204.

Danny Hatters, courtesy of Danny Hatters, p. 205.

Gregory Moseley, courtesy of Gregory Moseley, p. 206.

Bruce Grant, courtesy of Bruce Grant, p. 208.

Michael Tuohy, courtesy of John Morley, p. 209.

Martin Greasley, courtesy of John Morley, p. 211.

Timothy Anning, courtesy of Tim Anning, p. 213.

Jacqueline Munro-Smith, courtesy of Jacquie Munro-Smith, p. 215.

Richard Wettenhall, Bill Sawyer and Gerhard Schreiber, courtesy of John Morley, p. 216.

Mary-Jane Gething, Lloyd Finch, William Sawyer and Kenwyn Gayler, courtesy of John Morley, p. 218.

Kenwyn Gayler and Bruce Livett, courtesy of Jacquie Munro-Smith, p. 223.

Bruce Livett, courtesy of Fairfaxphotos/Simon Schluter, p. 225.

Geoffrey Howlett, courtesy of Geoffrey Howlett, p. 226.

Heung-Chin Cheng, courtesy of Cheng, p. 227.

Malcolm McConville, courtesy of Malcolm McConville, p. 229.

Paul Gooley, courtesy of Paul Gooley, p. 230.

Paul Gleeson, courtesy of Paul Gleeson, p. 232.

Ian van Driel, courtesy of Ian van Driel, p. 234.

Andrew Hill, by Irene Stanley, p. 235.

Marie Bogoyevitch, courtesy of Marie Bogoyevitch, p. 237.

Dominic Ng, courtesy of Dominic Ng, p. 238.

Stuart Ralph, courtesy of Stuart Ralph, p. 239.

Michael Griffin, courtesy of Mike Griffin, p. 241.

Leann Tilley, courtesy of Leann Tilley, p. 243.

José Villadangos, courtesy José Villadangos, p. 245.

Kathryn Holt, courtesy of Kathryn Holt, p. 246.

Diana Stojanovski, courtesy of Diana Stojanovski, p. 248.

Bio21, courtesy of Bio21 Molecular Science and Biotechnology Institute, p. 249.

Terrence Mulhern, Heung-Chin Cheng, Paul Gooley, Ken Gayler and Bruce Livett, courtesy of Beverley Bencina, p. 252.

Shayne Bellingham, courtesy of Lesley Cheng, p. 255.

Matthew Dixon, courtesy of Matt Dixon, p. 256.

Cheryl Chia, courtesy of Cheryl Chia, p. 257.

Index

THE MIEGUNYAH PRESS

This book was designed by Patrick Cannon
The text was typeset by Patrick Cannon
The text was set in 11 point Minion with 15 points of leading
The text is printed on 128 gsm Matt Art

This book was edited by Paul Smitz

THE
MIEGUNYAH
PRESS